Henry F. A Pratt

Astronomical Investigations

The cosmical relations of the revolution of the lunar apsides. Oceanic tides

Henry F. A Pratt

Astronomical Investigations
The cosmical relations of the revolution of the lunar apsides. Oceanic tides

ISBN/EAN: 9783337034399

Printed in Europe, USA, Canada, Australia, Japan

Cover: Foto ©berggeist007 / pixelio.de

More available books at **www.hansebooks.com**

Astronomical Investigations.

THE COSMICAL RELATIONS

OF THE

REVOLUTION OF THE LUNAR APSIDES.

OCEANIC TIDES.

BY

HENRY F. A. PRATT, M.D.

LONDON:

JOHN CHURCHILL AND SONS, NEW BURLINGTON STREET.

MDCCCLXV.

PREFACE.

When Pilate asked the question, What is truth? the apocryphal gospel of Nicodemus affirms that he was told in reply, that—Truth is from heaven: but that it is on earth, among those who, when they have the power of judgment, are governed by truth, and form right judgment.

Three conditions are thus declared to be necessary to the enunciation of the truth in all cases: the power of judging, the acceptance of true evidence, and the formation of a right judgment.

In Astronomical Science the combination of these conditions is very rare, for the power, when unconsciously severed from the will, fails to recognise the value of the evidence; or else, comprehending its force, permits it to remain unverified, as though valueless; the sense of the imperative duty of forming the right judgment, when the true evidence is presented, being only too often wanting, here as elsewhere, in those who have the power of judging, but not the courage to face the consequences of their judgment: for, perhaps unfortunately, those with whom, by an almost universal sufferance, if not necessarily, the power of judging is permitted to rest in this science, are parties in any issue that may be raised in which accepted theories are controverted, and thus are virtually made the judges in their own cause; for the unlearned public admits its incompetence to decide in such cases; but, as a consequence of this special interest, they almost unavoidably read all evidence that may be submitted to them through the distorting lens of their own preconceived opinions.

In the judgment of facts, however, on which observation can be brought to bear, and where prejudice must yield to the determining and determined results of a rigid scrutiny, they will, themselves, become truthful witnesses when their attention is sufficiently roused to the bearings of any given fact to the verification of which they are invited. The Author, therefore, now appeals to them, the practical astronomers, who are also through the force of circumstances the judges in his cause, to test the value of his views as to the true cosmical relations of the revolution of the lunar apsides, and the teachings that can be drawn from tidal phenomena considered in their oceanic origin; confident that if they can be once led to see the importance, not to say the necessity, of reconsidering accepted theories on these points, through a new and possibly true bearing of the actually observed facts,—the true evidence on which their judgment ought to be based, the right judgment will come, and these Astronomical Investigations not have been written in vain.

THE COSMICAL RELATIONS

OF THE

REVOLUTION OF THE LUNAR APSIDES.

INTRODUCTION.

The following paper on "The Cosmical Relations of the Revolution of the Lunar Apsides" was written at the suggestion of a friend, who considered the author's astronomical opinions of sufficient importance to deserve a formal discussion, and therefore kindly offered to communicate them to the Royal Society. It was accordingly placed in the hands of one of the secretaries of that Society, but returned by him as not suitable for the consideration of its members.

Perhaps it was declined because an attempt has been made in it to free an abstruse and difficult subject from the technicalities and obscurities in which it is commonly involved, by reducing the principles through which some of the phenomena resulting from revolving motion, combined with a gravitating attraction in a complex system, ought to be interpreted, to simple geometrical demonstrations, stripped of algebraic formulæ and symbolic calculations: though it may have been because the views advanced in it do not agree with the opinions of the day.

However this may be, the author regrets that an opportunity for discussion has been denied to him, for he conceives the whole subject to be of sufficient importance to deserve examination, notwithstanding any *seeming* unscientific method in the manner in which it is approached; while, if learned societies decline to discuss unaccepted opinions, either because they are opposed to those held by the greater part of their members, or else fear that they may compromise themselves as a body, by seeming for a time to endorse views which may turn out to be incorrect, but can only be tested by such a searching inquiry as it is in *their* power to institute, it becomes evident that they have fallen away from the end proposed in their institution, and, instead of aiding, render themselves real impediments to the advance of science.

The author still desires that his opinions may be examined, in order, if they are incorrect, that their fallacy may be established: he therefore now publishes them, but in the form in which it was proposed to submit them to the Royal Society, in order that his readers may have exactly the same opportunity of judging their merits.

The progress of science is unfortunately beset by two classes of difficulties, each of them presenting peculiar impediments to the onward movement. The first of these accompanies the observation and determination of phenomena and facts, and is sometimes so impenetrable that a series of successive patient observers has been necessary in order to establish a single relation.* The second lurks in the path of the interpreter of observed and determined phenomena and facts, and often leads him to give a false value to the results of observation.†

As the intrinsic worth of facts is measured, not only by the difficulty with which the knowledge of them has been gained, but also by the importance of the lesson which they appear to teach, there is always a tendency in the

* Until the time of Copernicus, the earth was supposed to be stationary. What a long period and series of observers passed away before its actual motion was determined! That is, under the present order of science.

† As in the case of Bradley, who discovering the apparent, interpreted it as an actual nutation of the earth's polar axis, and attributed it to the influence of attraction acting upon the assumed preponderating equatorial mass of the earth.

mind of the successful observer to estimate his facts (or supposed facts), by the labour which their attainment has cost him, and to interpret them more or less through his own preconceived opinions—hence there is always a possibility, nay, perhaps even a strong probability, that every newly ascertained fact will be at first misinterpreted.*

Even in those cases in which facts have not been overestimated, an importance is often given to the opinion of their discoverer in proportion to their actual value, so that he is raised into an authority on the subject, and his interpretation considered as certain as the fact which he interprets—all other methods of interpretation being most probably overlooked; so that it is possible that simultaneously with the knowledge of fresh facts a false method of interpretation may arise and be perpetuated for an indefinite time, the authority of a name being thus rendered an obstacle to the discovery of the truth.†

The science of Astronomy has been in all ages peculiarly exposed to this influence. Its facts are always learnt by slow processes and with very great difficulty, while they are, at the same time, so important, that every discoverer becomes an authoritative exponent of opinion; and yet, unfortunately, in order to become a discoverer a peculiar training is necessary, which *perforce* involves the adoption of preconceived opinions, so that, a given theory having once

* This was done by Galileo, who, when he discovered that balls of different weights appeared to take the same time in falling from the same height to the earth, rashly generalized that gravity acted equally upon all masses of matter at the same distance, forgetting that his experiments were on so small a scale, *as compared with the mass of the earth*, that differences could not be made perceptible to the senses of man, which nevertheless reason and philosophy show must exist. See *On Eccentric and Centric Force*, Part I., sec. 1, and Part III., sec. 10.

† This is strikingly seen in the case of Sir Isaac Newton, who, recognising the cosmical influence of gravity as a source of attraction, rashly generalized the principle that it was the source of the motions of the heavenly bodies, instead of a tendency in their constituent matter to draw them together, and bring them to a state of union and rest.

been accepted, every succeeding phenomenon will be of necessity read and interpreted through that theory.*

In this science, from the earliest ages, but certainly from the time of Copernicus downwards, the almost constant obstacle to onward progress has been the misinterpretation of observed phenomena. Every fact added to the general stock, when once it has been established, is an actual gain. The whole group of recognised facts forms the common store from which all interpreters must draw the bases of their inductions; but, though the facts are (perhaps) admitted by all, and even beyond the reach of question, any method of interpretation may be far from being so. And therefore, and as long as a possibility of error exists, every new system of interpretation is entitled to a hearing, and those who withhold this right expose themselves to the risk of opposing the advance of that truth which every inquirer is in reality labouring to define and put within the reach of all.†

The writer of these pages, after a patient and careful examination of the whole subject, has come to the conclusion,

* The predicted existence and subsequent discovery of the planet Neptune, through the perturbations caused by its mass, is given as a triumphant proof of the truth of the theory of universal gravitation. In reality it only shows that an attraction is seated in matter which acts at immense distances, and, when acting excentrically, alternately accelerates and retards the motion of those bodies on which it is acting, and causes relatively slight deviations from their true paths. It gives no further proof of the truth of the Newtonian interpretation or theory of universal gravitation than does the calculation and prediction of eclipses, which is quite independent of that theory.

† The theory of universal gravitation was constructed on a basis of misinterpreted facts, and then became a groundwork of preconceived opinions through which a misinterpretation of subsequently discovered facts was provided for and perpetuated. Thus, Sir Isaac Newton read, in the apparent, an actual variability in the moon's motion, and attributed this assumed variability to the attraction of the earth; and then, observing the effects of attraction in planetary perturbations and the tides, promulgated a scheme which, when once it was accepted, became the source of all subsequent interpretations; so that already recognised phenomena, as recession and precession, as well as every fresh discovery, such as the apparent nutations of the earth's polar axis, were interpreted by and read as further proofs of its accuracy.

that by studying the mutual relations of the whole series of observed phenomena from a new point of view, in which they are combined into groups, each of which is referred to certain distinct causes, principles can be developed more philosophic in themselves, and at the same time more easily subjective to examination and verification, and more facile of comprehension, and results elicited from these principles, under which far greater scope is given to the present limited range of astronomical theory.

The present range of astronomical vision, as far as the system of which the earth is a component part is concerned, is centred in the sun, to which motion has at length been attributed, and even the direction of that motion conjectured; so that the picture presented to the mind's eye is that of the solar system hurrying through space, and possibly revolving in an extensive orbit, the relations, range, and period of which are not even attempted to be surmised.

Grand as this ideal picture may be, and perfect, as far as it goes, it is yet a very limited one, when contrasted with the results which can be drawn from a more careful examination of the mutual relations of the observed phenomena; for, under such an examination, it can be shown that the solar system, vast and complicated though it be, is but a single member of a wonderful compound system, the general scheme and relations of which can be depicted: thus, the solar system is hurrying round an as yet unrecognised body or central sun*—which is itself again hurrying round an also unrecognised body or centric sun†—each of these bodies lying on what may for convenience be termed the plane of the zodiac, though in reality the actual planes of the orbits of motion are inclined to each other.‡

* See *On Orbital Motion*, by the author. (London: John Churchill and Sons.) Par. 12.

† This body is not mentioned in *Orbital Motion*. The author has only demonstrated its existence since the publication of that work.

‡ See *Orbital Motion*, Notes 96 and 117.

Moreover, the actual measures of the respective periods of these several bodies exist; for the lunar cycle (of recession) is the measure of the time occupied by the sun in revolving in its orbit, while the terrestrial cycle (of precession) is the measure of the period occupied by the central sun in its revolution.*

But the manner in which the phenomena can be made to speak does not end here; on the contrary, it can be shown that the centric sun is itself in motion—nay, more than this, although so far its period cannot be determined, the direction in which it is moving can be ascertained, when singularly enough it is found that the observed motion at present attributed to the sun is that of the centric sun, which, instead of revolving on the plane of the zodiac, like the component members of its system, is passing through (or across) that plane.†

In order to understand these several relations, it is necessary to regard the whole compound system as spheroidal in character; and to consider all these motions as taking place on the surface of the assumed spheroid—but with this difference, that the orbit of the centric sun is a great circle of the spheroid, while the other bodies are revolving on its surface; so that, while the plane of motion of the centric sun is through the centre of the spheroid, the general direction of the planes of the orbits of the other bodies is parallel to a segment of its surface; in consequence of which the plane of the orbit of the centric sun may be said in general terms to be at right

* See *Orbital Motion*, Notes 25 and 81.

† See Fig. 1. See also *Orbital Motion*, Fig. 1, where the path of the central sun should be treated as that of the centric sun. The evidence of the motion of the centric sun is drawn from the difference in period between the revolution of the terrestrial apsides and equinoctial points. The great difference between the periods of these two simultaneous revolutions in opposite directions, is owing to the path of the centric sun being across the zodiac.

angles to the common plane of motion of the members of its system.*

The centric sun is in reality revolving round a point or body situated within, and probably (relatively) in the neighbourhood of the centre of the spheroidal system—but still excentric in the path of the centric sun.†

This focal body of the orbit of the centric sun has peculiar relations to all of the bodies forming the entire system. It lies upon and is the common focus of the axes of their poles in the North Celestial Pole, and thus can be very properly termed the Celestial Polar Centre. It is an attracting body. It has drawn the bulk of the solid matter of the earth into the northern hemisphere. It is the source of attraction of the magnetic needle; and by its attraction maintains the stability of the polar axis of the earth and of the other members of the system. Its attraction varies with the distance through which it acts; and it is this varying attraction, acting upon the earth by slow degrees, through long periods of time or cycles, that has produced a large proportion of the physical changes in the constitution of the earth; and besides this, it is probably the primary source of the electrical and magnetic currents which pervade the whole system, and become, from time to time, visibly sensible in the Northern Lights.‡

Three simple but well-recognised principles, acting through the relations determined by the spheroidal system just described, when rightly and carefully applied to the interpretation of the observed phenomena, at once throw a fresh

* The simplest way of studying these relations is to view the centric sun as moving in a direction parallel to the mean polar axis, or from north to south, while the other bodies are passing respectively from west to east.

† This excentricity will have a double aspect; for though it may not be great with reference to the volume of the imaginary spheroid, it would yet, relatively to the volume and mass of the earth, cause immense differences in its actual distance in the different and ever-varying positions of the entire system and its relative parts.

‡ See *Orbital Motion*, Note 95.

flood of light upon the physical history of the earth. These principles are—

First. The mutual attraction which the several bodies have for each other, and especially the increase which takes place in this attraction inversely as the distance diminishes, and *vice versâ.**

Second. The inclination of the planes of the several orbits to each other and to the equators of their proper bodies; † and,

Third. The excentricity of the several focal bodies.‡

The source of the first of these principles is, probably, an inherent tendency in all matter to combination, union,

* This attraction is twofold in its character: between a primary and its satellites, when it acts concentrically, resisting the force which maintains the persistent motions of the satellites round it, and tending to draw them into itself, thus becoming a true centric force, as in the case of Jupiter and its moons, a due balance between this attraction and the active motive force maintaining the systemic relations; and between any two (or more) other bodies, as between Jupiter and the earth, when it acts eccentrically, tending to disturb the systemic relations, and produce what are termed perturbations. The Newtonian theory has failed to discern the force of this distinction.

† This inclination is the cause of one form of variation in the degree of attraction—that acting through the polar axis, since it causes a regular alternate variation in the distance between the revolving body and the celestial polar centre.

There are three ways in which orbital inclination might originate:—

1. Through the eccentric attraction of a remote concentric body, as explained in *Orbital Motion*, Note 117, and Fig. 13.

2. Through alternate systemic concentration and expansion, also indicated in *Orbital Motion*.

3. During the progressive motion of the centric sun across the zodiac, the inclination of the zodiacal plane would diminish as its centre was approached, and increase again after it was passed.

The two former appear to be both of them in operation,—the one acting on the plane of the moon's orbit, the other on that of the earth.

‡ This provides for another form of variation in the degree of attraction. The probable cause of excentricity has been discussed in *Eccentric and Centric Force* and *Orbital Motion*. There is another way of accounting for it: by regarding each revolution as an act of projection, under which circumstances the law of projection given by the author in the works just referred to would be followed, and the revolving body increase its velocity of motion as it increased its distance from the focal body of its orbit.

rest;* while the second and third appear to depend upon the varied modifications and modifying action of the first, when resolved into the several forces drawn from the mutually re-acting bodies, and their antagonistic relations to the active force through which the persistent motions of the universe are maintained.† However this may be, it is sufficient to regard them here as principles drawn from the actual and observed phenomena, when, read in and through them, a wonderful provision is at once found for all the changing relations of the whole system.

Thus, to apply them to the earth considered as a spheroid of revolution in equilibrium.‡ The bulk of the land is in

Perhaps both forms are in actual cosmical operation, the excentricity of the celestial polar centre depending upon the motion of the centric sun following the laws of projection; while the secondary and tertiary, or more remote forms, result from the simultaneous motion of both primary and secondary alternately towards and from each other. Two causes seem to be in operation, determining variations in the planes of orbit—that re-acting upon the moon's path (*Orbital Motion*, Note 117), and that acting upon the ecliptic (*Orbital Motion*, Note 119); it is, therefore, not unlikely that two causes, resulting in two forms of excentricity, primary and subordinate, are simultaneously acting and producing the combined result.

* See *Eccentric and Centric Force*, Part I., sec. 3, and Part III., sec. 10; and *Orbital Motion*, Chap. III., par. 112 and 113.

† See *Eccentric and Centric Force*, Part IV., sec. 7; and *Orbital Motion*, par. 165.

‡ In a spheroid of revolution of equal density, the centre of gravity is on and at the centre of the axis of revolution; so that radii drawn from the same latitude to its centre of gravity, or any point of its polar axis, will be of equal length. In a spheroid of revolution of unequal density, the centre of gravity will still be on the axis of revolution; but radii drawn to it from the same latitude will be longer in the less dense than the more dense portions, equilibrium being preserved in the parallel planes of revolution by an increase in volume on the one side compensating for an increase in density on the other.

It must be remembered, that in a spheroid of revolution, there will be a physical tendency to elongate the polar axis at one set of velocities, just as there may be a physical tendency to shorten it at another. Thus, if a body like the earth is revolving with a low velocity of revolution with regard to its axis of motion, determined by the number of revolutions accomplished in a given time, but with a high velocity of transition, as between its equatorial surface and space, its equatorial fluid particles

the northern hemisphere; and of the water in the southern. The land, as a whole, has a greater specific gravity than the water: hence the centre of gravity of the earth will be north of the equator; and, as it is in revolution, on its polar axis. But since the earth is necessarily in equilibrium, a plane parallel to its equator, and drawn through its centre of gravity, would divide it into two portions of equal weight, but unequal volume; for the polar axis is divided by it into two unequal parts, the shortest of which is in the northern hemisphere, the centre of figure being in the southern or larger portion; so that the greatest volume of matter lies south of the plane drawn through the centre of gravity. On the other hand, the equatorial plane may be considered to pass through the centre of figure, and so give an approximate equality in volume to the two hemispheres. But now a preponderance of weight is found in the northern hemisphere,—it is heavier than the southern: hence the centre of figure becomes a true centre of suspension—the polar axis the line of suspension, the celestial polar centre the centre of attraction towards which the suspension is directed. The relations of the earth, primarily considered, are thus not unlike those of a balloon; its greater volume, but diminished density, being in the southern hemisphere—the suspending power that force which maintains the movements of the whole system.

Under this view, increase in volume is used revulsively to compensate for diminution in density, in order to preserve individual equilibrium, and with it the stability of the polar

will recede upon its axial motion, and seek a diminished velocity in the polar regions, and thus tend to elongate the polar axis. To develop centrifugal force, it is necessary that the velocity of revolution, as well as that of transition, should be high, and that there should be free scope for the action of the antagonistic forces. The latter condition could probably only be attained in a spheroid of revolution, the whole of whose surface consisted of fluid particles. That centrifugal force is not in action at the earth's equator is demonstrated by the equatorial currents, whether of the ocean or atmosphere. But if it is not in action, then another reason is given for affirming that the earth's polar axis is its longest diameter. See *Orbital Motion*.

axis. The same principle holds good on the several planes of revolution; so that, if a plane be taken through the polar axis, the hemisphere containing the most land will be opposed by an hemisphere containing most water,—that is to say, the water line will be higher in proportion as its surface area increases, the radius of volume drawn from the centre of figure longer in that hemisphere where there is least land.*

Owing to this primary difference in the relative densities of land and water, and the fluid nature of the more voluminous and less dense element—water—variations in the equilibrium of the earth, from whatever cause they may originate, will be necessarily accompanied by a variation in the level of the water line; the law of this variation being, that under the influence of a disturbing cause the water will flow to that point whose relative density has diminished, from the point where it has been increased.†

This is the primary law of revulsion in equilibrium. Its tendency when called into action will be to cause a tide to pass from one hemisphere into the other, and it is in reality a modification of the law under which gases recede from the centre of the earth.

In considering the action of this law upon the earth in its cosmical relations, it must be remembered that increase in

* This is strikingly illustrated on the eastern shore of the Pacific Ocean, where the water maintains a higher level than on the western Atlantic; although physical causes, taken by themselves, would make the Atlantic the highest.

† The same law holds good when the relative gravity of either hemisphere (divided by a plane through the polar axis) is disturbed in any other way, as by an addition to its volume; so that, if a wave were generated on the equator at any point, there would be a tendency to generate a similar wave at the opposite side of the equator; and thus two waves would result, 180° apart, the latter originating in the law of revulsion, in order that the stability of the earth's axis as a spheroid of revolution may be maintained. This revulsive law will cause any two waves travelling round the earth, to whatever cause they may owe their commencement, to preserve an interval of 180°, and therefore, in a great measure, regulate the velocity, and preserve the synchronism of the tides. (See *Oceanic Tides*, by the author.)

gravity is only another way of expressing *increased subjection to the influence of an attracting force;* so that, *inter alia,* the same particles of matter would be heavier at the North Pole than at the South Pole, because at the North Pole they would be nearer to, and therefore more subjected to the influence of the celestial polar centre.

It results from this, that the land in the northern hemisphere is, from position, relatively heavier than that in the southern, as well as indeed the water;* hence from this cause there will be an increased revulsive effect in the southern hemisphere, so that its volume in figure will be greater than that of the northern.† Besides this, owing to this combined form of attraction and revulsion acting through the polar axis of the earth, there will be a tendency to elongate that axis;‡ so that the earth's true figure will

* Even with reference to the centre of gravity of the earth, the same particles of surface matter will be heavier in the northern than in the southern hemisphere in the same latitude, because the distance from the centre of gravity increases in passing to the south, owing to the excentricity of the earth's centre of gravity.

† It is through this revulsive action that the *ovoid* form of the earth results, the broad end of this figure being thus placed in the southern hemisphere.

‡ The relations of the polar axis and equatorial diameter are inverse to each other. As the one elongates, the other contracts. A provision is found for this alternation in the ocean beds, more particularly that of the Atlantic, which, as a longitudinal indentation in the earth's crust, becomes deeper and narrower as the polar axis is elongated, thus causing the contraction of the equatorial diameter.

Some of the phenomena which appear to contradict the revulsive theory are caused by these relations; for, as a consequence of the subsidence of the bed of the Atlantic, the coast of Greenland will be slowly submerged,—dragged down by the sinking ocean bottom,—while the lateral or corrugating tension will often reach a point at which the crust of the earth will crack and recoil, some of the strata yielding. In this manner, sudden elevations will occur, often of very considerable extent. These will be more liable to happen at the margins of the ocean, and in the neighbourhood of the tropics. For the same reason, in approaching the poles, these sudden releases from tension will progressively increase in ratio towards the poles. Every exceptional case will be found susceptible of explanation when carefully studied in all its relations through sound views of physics.

approximately resemble that of an egg, having its base or broad end in the southern hemisphere.*

From variations in position on the earth's surface, the effects of variation in the position of the earth in space have now to be considered.

If it were possible that by very slow degrees, extended over long periods of time, the earth was alternately drawing near to and receding from the celestial polar centre, it would follow, as a necessary result, that a cyclical revulsive tidal wave, co-extensive in its volume with the degree of variation in distance, would be alternately passing from one hemisphere to the other, which would be read on the surface of the earth as a variation in the level of the land.

But such a variation has taken place,—nay, is still progressively taking place on the surface of the earth; indeed, the evidence of it—the alternate oscillations in its level—is a primary means of distinguishing between the great geological epochs.

* Sir Isaac Newton assumed that the figure of the earth was that of an oblate spheroid on purely theoretical grounds. This figure has never been demonstrated, and the geometrical results drawn from actual measurement are against it; for arcs of the meridian become longer as they approach the poles, which at once proves that the earth's polar axis is the longest. Astronomers get out of this difficulty, in order to maintain their theory, by affirming that this elongation of the arcs of the meridian is due to the fact that gravity acts vertically from the surface of the earth; but in affirming this, they in reality give up the Newtonian theory altogether; for it refers gravity to the centre of gravity, irrespective of the earth's figure. The fact is, that the direction of gravity, with reference to the surface of the earth, cannot be determined experimentally; for the spirit-level will always give an horizon at right angles to the plumb-line. But how can it be proved that this agrees with the level of the surface of the earth, or the horizon of oceanic water at rest? This is an assumption, and, in all deviations from the perfect sphere, an unphilosophical one. In a spheroid, the horizon of the spirit-level (or of isolated surfaces of still water) will only agree with the horizon of figure at the poles and the equator (or its equivalent). But if this is so, then no correction is to be made for differences in the direction of gravity in calculating the true arcs of the meridian; and their progressive increase from the equator to the poles is a positive demonstration that the earth's polar axis is longer than its equatorial diameter.

If it were further possible that these alternate variations in level could also resolve themselves into variations in degree,—so that during one series of the cycles of change the earth was drawing successively nearer and nearer to the celestial polar centre, and during another series of such cycles receding further and further from it,—then the revulsive tidal wave would be itself a variable one, increasing in volume during its *serial* recession from, and diminishing in volume during its *serial* re-approximation to, the celestial polar centre.*

But the actual oscillations of the earth's surface level have varied considerably in degree according to the results drawn from geological research; the degree of alternate subsidence and re-elevation at any given point having been much more extensive on some occasions than on others.

Now, these varying relations are not only accounted for, but rendered necessary by the theoretical extension of the astronomical system here advocated; for, if the central sun is passing round the centric sun in an orbit inclined to its

* At first sight this appears to be a contradiction to the principle of revulsion laid down. It is, however, only a modification of that principle, which can be easily made intelligible; for it is owing to the inclination of the orbit of the central sun that this cyclical tidal wave exists. Now this inclination is a variable one, attaining its maximum when the centric sun is most remote from the celestial polar centre, and reaching zero at its point of greatest proximity; so that the actual wave (or alteration in the level of the land) will be the greatest when the whole system is furthest from the celestial polar centre, although the land will attain its maximum of elevation, simultaneously with a minimum tidal wave, when the whole system is nearest to the celestial polar centre.

These seemingly incongruous relations arise from the blending of two orders of waves,—that dependent on the revolution of the centric sun, under which the water has made its greatest recession from the circumpolar regions (especially the northern), when nearest to the celestial polar centre; and that dependent on the revolution of the central sun, which, since it is due to the inclination of its orbit, sinks to a zero point at that period, but attains its maximum when the centric sun's tide has returned to the circumpolar regions: so that the gross result is that the motion of the centric sun makes a slow but determined serial change in the degree of the alternations in the level of the land caused by the revolution of the central sun.

equator, it will then be alternately drawing near to and receding from the celestial polar centre, with its whole system;* while further, if the celestial polar centre holds an excentric position in the orbit of the centric sun, it follows, as a matter of course, that the centric sun in its revolution carries its whole compound system alternately nearer to and further from the celestial polar centre:† so that, the cycle of precession being taken as the cycle of the central sun's period (the centric sun's path, of course, representing an unknown series of such cycles), during one-half cycle it would be drawing near to the celestial polar centre,‡ and the revulsive tidal wave would be ebbing from the northern hemisphere to the southern, and from the polar regions to the equatorial; the effect on the surface of the earth being that the land forming its equatorial belt would be slowly subsiding into the retreating waters, that in its circumpolar regions, especially the northern, as gradually emerging from them; while during the next half cycle it would be receding from that centre,§ and the tidal wave would now be passing from the southern hemisphere to the northern, or ebbing in the equatorial regions and flowing in the polar; the land now rising in the equatorial and subsiding in the circumpolar regions.|| Moreover, as during succeeding cycles in precession, the centric sun would be advancing in its path, or moving round the celestial polar centre, and at the same time varying its distance from it, it is evident that these alternations must vary in degree, and the tidal wave be the greatest—the oscillations in the level reach their

* Fig. 2. † *Ibid.*

‡ Or passing from its point of greatest remoteness from, to that of greatest proximity to, the centric sun. Its point of greatest proximity to the centric sun will be its point of greatest proximity to the celestial polar centre in that revolution.

§ Or passing from its point of greatest proximity to, to that of greatest remoteness from, the centric sun.

|| See *Orbital Motion*, Notes 22, 95, and 268.

maximum when the centric sun was furthest from the celestial polar centre.*

Another consequence flows from these changing relations; for, in drawing near to the celestial polar centre, an actual concentration with a diminution in excentricity takes place in the solar system;† while increase in proximity to that centre enables it to receive renewed force or energy from it.‡

These, which are only some of the first and most prominent results drawn from the present attempt at a re-interpretation of well-recognised astronomical phenomena, have this peculiar feature, that while grouping the phenomena themselves into classes, and then tracing each class to its direct cause, and so drawing the outline of a more perfect system than any that has yet been proposed, they fulfil this further condition of such a system, that they indicate the astronomical measure of the cosmical changes which have taken place, and are taking place on the earth's surface—pointing to the cycle of precession as the great geological "day," or revolution—and further indicating the source of variation, in the effects produced during that cycle.§

Such an astronomical measure must exist, seeing that from whatever cause variations in the earth's individual relations may originate, their nature proves that they are cosmical in character, and, therefore, responsive to a cosmical cause which permeates the whole system, of which the earth is a relatively remote member: hence every system which does not point to it is inadequate to the end for which it is proposed. Astronomy must furnish the link which binds the sciences together, and unravels some of their obscurities—providing the basis through which all their relations are to be read, their relative

* See Fig. 2.

† See *Orbital Motion*, Notes 52, 96, 119, 138, and 173; and *Eccentric and Centric Force*, Part IV., sects. 1, 4, and 7.

‡ See *Orbital Motion*.

§ See *Ibid.*, Note 268.

quantities determined; and, above all, the geologist, the zoologist, and the botanist must turn to the astronomer for the clue to possibly missing links in the chain of their science.

The interpretation and grouping of astronomical phenomena is very simple.

1. The moon passes round the earth from west to east; its passage causing the sidereal heavens, including the sun, to appear to pass from east to west.

2. The visible hemisphere of the moon gravitates to the centre of gravity of the terrestrial system, in consequence of which the sidereal heavens pass round its circumference as it passes round the earth, and give it an apparent revolution on its axis.

3. The earth revolves upon its polar axis from west to east; the passage of a given meridian round that axis causing the sidereal heavens, including both sun and moon, to appear to pass round it from east to west.

These are the individual motions of the terrestrial system.

4. The terrestrial system passes round the sun from west to east, its passage causing the sidereal heavens to appear to pass round it from east to west; while, owing to that passage, the sun appears to move through the signs of the zodiac from west to east.*

The difference between the sidereal and synodic lunar

* Three distinct phenomena singly and simultaneously furnish the measure of the period occupied by the terrestrial system in passing round the sun.

1. The sun passes through all the signs of the zodiac, and returns to its point of departure, thus making a complete apparent revolution in space, which is the reflection of the actual revolution of the earth.

2. The same star returns to the meridian with the sun—that is to say, the solar and sidereal meridian once more coincide as measures of time—the earth having in the interim made one more revolution on its axis with reference to the sidereal heavens than with regard to the sun; and

3. The sidereal and synodic points of departure of the lunar revolution once more coincide, the moon having in the interim made one more revolution round the earth, with reference to the stars than with regard to the sun.

month and the sidereal and solar day is caused by the progressive motion of the earth.

5. The sun* moves round the central sun from west to east. This passage causes the point at which the earth's annual revolution round the sun is completed, to recede from the central sun from east to west;† which recession is read in the heavens by a passage of the zodiac from west to east round the lunar node; so that the complete recession of that node is the measure of the period of the revolution of the solar system in space.‡

This is accompanied by an advance of the lunar apsides, which pass *once* through the signs of the zodiac, but *twice* from conjunction to opposition, in (approximately) the same period, depending on the attraction of the central sun;§ by an oscillation or variation in inclination of the moon's path, caused by a variation in that attraction, as the earth alternately draws near to and recedes from that body;‖ by an apparent nutation of the polar axis of the earth, determined by the actual stability of that axis or fidelity with which it points to the celestial polar centre, in consequence of which it appears to nutate with reference to any intermediate star;¶ and, further, by a progressive variation in the length of the lunar synodic revolution.**

These phenomena constitute a single cycle to which the name "lunar cycle" may, for convenience, be given, since its most readily interpreted unit is the lunar recession. It is determined by the revolution of the sun, and is the measure of the period of that revolution.†† It resolves itself into two groups: the first and shortest comprehends the apsidal

* The sun revolves upon its axis, and the planetary systems revolve round it, with their individual and systemic motions; but these can be passed over here.

† See *Orbital Motion*, Fig. 3.

‡ See *Ibid.*, Notes 25 and 170.

§ See *Ibid.*, Note 167.

‖ See *Ibid.*, Note 117.

¶ See *Ibid.*, Note 79.

** See *Ibid.*, Fig. 20.

†† See *Ibid.*, par. 48.

revolution and oscillation in plane of the moon's orbit; the second, the complete recession of the nodes, the revolution in nutation, and the variation in the lunar synodic period,—the difference in period occupied by these two being caused by, and a primary measure of, the motion of the central sun in space.*

6. The central sun moves round the centric sun from west to east. This passage causes the earth's equinoctial points to recede round the zodiac from east to west. This is known as the precession of the equinoxes; a complete revolution in precession being the measure of the central sun's period.†

It is accompanied by an advance in the terrestrial apsides,‡ by a variation in the obliquity of the ecliptic,§ by an apparent nutation of the earth's polar axis,‖ and by a progressive variation in the length of the tropical¶ year,—these phenomena constituting the terrestrial cycle.** They are parallel in their teaching value to those in the lunar cycle, and, like the latter, are divisible into two groups. The first and shortest includes the apsidal revolution and variation in the obliquity of the ecliptic; the second, the precession of the equinoctial point, the apparent nutation, and the variation in length of the tropical year. The difference in the period of these two groups is caused by the motion of the centric sun, and is of such a great proportionate extent, because that body is passing *across*, and not revolving within, the zodiac.††

7. The centric sun is passing round the celestial polar centre from north to south, on the plane of its polar axis. This change in direction (from an approximate equatorial to an approximate polar plane) causes it to move across the zodiac, and has been hitherto read by astronomers as the proper motion of the sun.‡‡

* See *Orbital Motion*, Note 138.
† See *Ibid.*, Note 81.
‡ See *Ibid.*, Note 138.
§ See *Ibid.*, Notes 96 and 119.
‖ See *Ibid.*, Note 79.
¶ See *Ibid.*, Fig. 20 and Note 173.
** See *Ibid.*, par. 49.
†† See *Ibid.*, Notes 119 and 138.
‡‡ See *Ibid.*, Fig. 1. There treated as the orbit of the central sun.

These phenomena have been examined and discussed elsewhere; and the grounds for grouping them in the manner here indicated, carefully weighed.* The whole subject has been considered under a fresh aspect, and from an entirely new point of view, in the paper on "The Cosmical Value of the Revolution of the Lunar Apsides," and the strict coincidence in the results drawn from this double process amounts to an absolute demonstration of the accuracy of the view now advocated.†

It is of course desirable, in offering a new theory to the world, to give, where it is possible, some clear and simple test by which its real value can be subjected to a rigid scrutiny. When *Orbital Motion* was published, the author invited the Astronomer Royal and the practical astronomers of England to make a search for the central sun, whose bearings and relative position in space he ventured to indicate.‡ He now formally invites those gentlemen to verify his views of the true significance of the phenomenon known as apsidal advance. The test is a simple one. He affirms that the lunar perigee, in passing from conjunction to conjunction, only advances through half of the signs of the zodiac. He is not aware that this is a recognised fact. He offers it as a predication, and challenges its admission or disproval. He is willing to stake the truth of his whole theory upon the results of observation on this single point, and waits for the revelations of such observation without anxiety, fully conscious that they will agree with his prediction.

But if they do agree with his prediction—if it should be discovered and admitted that two revolutions of the lunar

* See *Orbital Motion*, passim.

† The proofs are drawn from one series of phenomena in *Orbital Motion*. They are now drawn from another and perfectly distinct set of relations, both of which demonstrate the same causes as being in operation.

‡ See letter dated 18th October, 1863, published in the *Times* newspaper

apse are necessary, in order that it may advance through all the signs of the zodiac, this position will have been reached, as a result of this crucial test, that apparently incompatible, not to say impossible, conditions are found to be coexistent—for how can the same relations make simultaneously an entire revolution in space when computed from one point, and only a half revolution when computed from another? And then the author is confident that there will be no escape from the dilemma in which astronomers will find themselves but the acceptance of his theory.

HEREFORD, *15th March*, 1865.

THE COSMICAL RELATIONS

OF THE

REVOLUTION OF THE LUNAR APSIDES.

Amongst the many interesting phases of motion which present themselves to the thoughtful observer in the study of physical astronomy, the revolution of the lunar apsides is by no means the least instructive. In studying this phenomenon, it is surprising to find how lightly it is touched upon, even by some of those who have elaborately discussed the whole field of astronomical change; for, beyond a dry description of the fact, and sometimes an attempt at a mechanical explanation of it, it is for the most part passed over in silence.

And yet, perhaps, this is not surprising, for in reality it presents a double anomaly under the received theory of gravitation; and hence it may still be regarded as an astronomical enigma, possibly containing the elements of an unsolved problem.

For these reasons it presents a fair field for investigation, and in a careful examination of its actual relations data may even be found for a more extended and comprehensive view of the whole subject.

The relations of the moon's orbit, known by astronomers

as its apsides, are indicated by an imaginary and purely theoretical line, which is supposed to pass through the relatively long axis of the moon's nearly circular path round the earth. They are determined by the excentric position of the earth in that orbit, under which the moon during its transit round varies its distance from the earth, alternately drawing near to and then receding from it, and are in fact assumed to lie upon, and hence are referred to a line passing at any given time from the point of the moon's greatest proximity to, through, and thence to the point of its greatest recession from the earth.

The designations of the points where the moon crosses the line of its apsides are *perigee* and *apogee* respectively; the perigeal point being, of course, that of greatest proximity.

The revolution of the lunar apsides, the probable cause of which is about to be considered, is this, that the perigeal and apogeal points of the moon's orbit do not represent fixed points in space,—that is to say, it is found under observation that their relations are not constant, whether to the sun on the one hand or the zodiac on the other, but progressively changing, for they are advancing in the order of the signs; so that, starting from conjunction with the sun in 4½ years, the moon's perigee is found to be in opposition to that orb; and in a second 4½ years, having still continued to advance, it will have returned to conjunction once more; and thus in every 9 years the lunar apse makes a complete revolution in space with reference to the sun.

The anomalies which this revolution presents are to the astronomer very clear—for,

1st. Its period is only 9 years; that of the recession of the lunar nodes being 18—while

2nd. It advances in space, whereas the lunar nodes recede, or move in an exactly opposite direction.

And it is on this account that a purely mechanical explanation of the phenomenon has been adopted, in order to separate it completely from the ordinarily recognised effects of the laws of gravitation.

On attentively considering this revolution, it is difficult to dissociate it from the recession of the lunar nodes, although not only have these two concurrent circumstances no phenomenon in common, but they even appear at first sight to be in all their relations opposed to each other, since it takes somewhat more than two periods of the one to complete the full period of the other, and they are moving in opposite directions.

In the relations of the half period of nodal recession to the whole period of apsidal revolution, the first link in a possible chain of evidence does, however, seem to present itself, as a guide to further inquiry. Another link can be drawn from an associated phenomenon, which more closely connects them, for the moon's orbit has not a permanent degree of obliquity to the ecliptic, but oscillates to and fro across it, a single oscillation exactly coinciding with one revolution of the apsides; a complete or to and fro oscillation being performed in somewhat less time than a complete recession of the lunar nodes.

Thus, the revolution of the lunar apse exactly coincides with the oscillation of the lunar orbit,—that is to say, during one revolution of the apse the orbit crosses the ecliptic in one direction, and in the next revolution recrosses it in the other, while both of these associated phenomena have a definite relation to the period of lunar nodal recession.

But the oscillation of the lunar orbit is evidently a single to and fro act, depending upon a definite cause which pro duces a regularly oscillating result.

Hence the first inference is, that a single revolution of the lunar apsides is in reality the measure of only half a phenomenon,—half of an actual revolution in space.

And yet that the lunar apse does make two revolutions during a single recession of the nodes cannot be denied.

Is it possible to reconcile such an evident discrepancy?

Perhaps an examination of the relations of the terrestrial apsides may throw some light upon it.

These, like the lunar apsides, are revolving in space, and, like them, are advancing in the order of the signs; but, as was to be expected from the difference in time occupied by a lunar and a terrestrial revolution, their period is greatly prolonged, the perihelion point of the earth taking some 21,000 years to pass through all the signs of the zodiac or make a complete circle in space.

But the advance of the terrestrial apse in space is not the only parallel relation between these two revolutions, for there is a precession of the terrestrial equinoxes which bears the same relation to the terrestrial orbit that the recession of its nodes does to the lunar; and then there is a progressive, though slow, variation in the obliquity of the ecliptic which parallels the oscillations of the lunar orbit; so that there are two sets of precisely similar phenomena, one of them affecting the terrestrial orbit, the other the lunar.

The question which now suggests itself is, What relations do each of the three phenomena of either set bear to the similar phenomena of the other?

It is a question easily answered; for, while the equinoxes pass round the zodiac against the order of the signs or retreat along them once in some 25,868 years, the terrestrial apsides pass round it in the order of the signs or advance along them once in some 21,000 years, while the ecliptic oscillates to and fro once in the same period as the apsides revolve.

The points of accord between the two sets of phenomena are thus at once seen to be, that the period of precession in the one, like that of recession in the other, is the longest, and so marks the longest revolution; while the period of oscillation of the ecliptic (to and fro) coincides with that of the revolution of the apsides; and, on the other hand, the points of difference are these, that the terrestrial apsides revolve but once during the oscillation of the ecliptic and retreat of the equinoxes, whereas the lunar apsides revolve twice during a single oscillation (to and fro) of the lunar orbit and recession of the nodes.

This difference in phenomena, which otherwise are so very similar as to appear to illustrate a general principle and depend on a similar cause, is very remarkable; and yet, after all, it is not so very unexpected, for it affirms what was suggested at the outset, that the revolution of the apsides is part and parcel of a single act to which the three several coincident phenomena belong.

But, if the revolution of the apsides illustrates, in combination with oscillation and recession, a single act, How is it that this act occupies two revolutions of the apse in the lunar orbit, and but one in the terrestrial?

How, indeed? In order to answer this question another must be asked, the answer to which will, perhaps, throw some light upon this remarkable difference. With reference to what is the revolution of the lunar apse computed? And when this has been replied to, it must be considered with reference to what the revolution of the terrestrial apsis is calculated.

Now, the revolution of the lunar apsis is computed from the sun—that is, from a departure from conjunction with, to a return to that relation to that orb; whereas the revolution of the terrestrial apsis is measured by a direct reference to the signs of the zodiac. This is an important difference, for it is possible that the lunar apsis might make an apparent revolution in space (from conjunction to conjunction), and yet only pass through half the signs of the zodiac in doing so.

To realize the possibility of this seeming paradox, a single assumption is required,—that the sun itself is describing an orbit in space round an excentric focal, and, as yet, undiscovered body or central sun, the period of which coincides with, and is measured by, what for convenience may be termed the lunar cycle.

If, for a moment, this could be conceded, the explanation would become very easy; for then

1st. The passage of the sun in its orbit would cause a retreat of the terrestrial system round that orbit, which would be recognised on the zodiac by the recession of the lunar nodes.

2nd. It would cause the central sun (were that body visible) to appear to advance in the order of the signs, and complete a full apparent revolution during the full revolution of the sun; the revolution of the sun being thus read in the skies by an apparent revolution of the central sun; and

3rd. As it approached the central sun the orbit of the moon would become gradually more inclined to its equator, crossing the ecliptic in doing so; while, as it receded from that body, the moon's orbit would become less inclined to its equator, and thus recross the ecliptic.

Admitting these three results as necessary consequences of the assumed revolution of the sun, another has now to be claimed; or,

4th. That the lunar apsides would always rest upon or point to the apparently advancing central sun; the moon at its perigee lying between the earth and the central sun.

This, of course, it would do, in virtue of the attraction of that body; but now, what would happen from these mixed relations? Why,

1st. That the central sun, as it advanced in the order of the signs of the zodiac, would carry the lunar apsides with it, *pari passu*.

2nd. That in doing this it would cross the retreating

terrestrial system, or receding node, twice in each revolution, at the opposite sides of the sun's orbit.

3rd. That each time that it crossed the lunar node, the terrestrial system then lying between it and the sun, the perigeal point of the moon would be in opposition.

4th. That each time that it was in quadrature with the points where it crossed the lunar node, the sun would lie between it and the terrestrial system, and the lunar perigeal point thus necessarily be brought into conjunction. Hence,

5th. That in passing from conjunction with the lunar node to quadrature with the point where that conjunction occurred (or actual opposition with the lunar node), the lunar perigeal point would be passing from opposition to conjunction; and,

6th. That in passing from opposition to the lunar node, to conjunction with it, the lunar perigeal point would be passing from conjunction back again to opposition.

That is to say, as the central sun passed from conjunction through opposition to conjunction again with the terrestrial system, the lunar perigeal point would make a single complete apparent revolution (with reference to the sun).

But this would occur twice during a single retrograde motion of the node, actual revolution of the sun, or apparent revolution of the central sun.

Hence two revolutions of the lunar apsides would necessarily coincide with a single recession of the lunar nodes.

Such an explanation as this of the revolution of the lunar apsides, would strictly accord with the theory of gravitation; for the attraction of the central sun would cause the apsidal line always to point towards it.

The possibility of such a theory as the one here offered being the true explanation of the form of motion under discussion, can hardly be denied. The manner in which it extends the received astronomical generalizations, and widens the field of investigation, must now be briefly considered, in order to determine how far it is a probable, as well as a possible way of accounting for a very interesting phenomenon.

The first consequence that flows from it is the change in the value which it gives to the phenomenon of recession; which amounts to a remodelling of opinions on this branch of the subject.

If the recession of the lunar nodes were the only form of recession recognised in practical astronomy, then the value now claimed for it might be called in question, because of the antecedent revolutions of the earth and moon, which ought to be accompanied by, and so occasion, similar results.

But these revolutions are accompanied by, and do occasion recession; and it is by the recessions which follow their motions, that these motions in space have been identified; but, owing to the recessions which they occasion differing in form, the actual identity in principle of their teaching value has been overlooked.

The first and simplest form of recession is that of the whole heavens round the earth during its axial revolution, upon which the difference between day and night depends. Similar to this is that of the heavens round the moon during its passage round the earth, perceived through its diurnal advance amongst the heavenly bodies.

The second form of recession is that recognised only after an interval of time, under which the whole aspect of the heavens differs, say in the winter, from its appearance in the summer; for here the heavenly bodies recede once round the earth and the sun during the annual revolution of the former, and occasion the difference between solar and sidereal time; owing to which the sidereal year comprises one more axial revolution of the earth than the solar year, though occupying exactly the same period of time—the effect of the progressive advance of the earth round being to lengthen its axial revolution with regard to the sun, by which process the sidereal revolution in excess is broken up into fragments, one of which is added to and included in each solar day; so that here the annual sidereal recession is occasioned by the progressive or orbital motion of the earth, and is confessedly a measure of its annual revolution round the sun.

The same form of recession is witnessed as the moon traverses its orbit, for its sidereal period is shorter than its synodic; so that it also performs one more revolution in each year, when computed from a fixed star, than when its course is measured from the sun; this sidereal revolution in excess being also, like the last mentioned, broken up into fragments, one of which is added to and included in each of its synodic revolutions,—this added revolution being thus a further measure of the earth's progressive motion round the sun in space.

On considering these several recessions, they are found to be measures of revolutions in space, exactly defining the periods of either of the revolutions indicated,—viz., the axial, and then the orbital, motion of the earth; and, finally, the orbital motion of the moon, their peculiarity being that the sidereal heavens recede (or move from east to west) round the revolving or advancing body.

In the recession of the lunar nodes, and the precession of the terrestrial equinoxes, this peculiarity is reversed; for here the sidereal heavens advance with reference to the retreating node or equinoctial point, or pass from the west to the east; so that if this change in the direction of the apparent motion can be accounted for, the inference becomes a demonstration that these recessions also are measures of the periods of revolution of other concentric revolving bodies.

And yet nothing is perhaps easier than to account, and that most satisfactorily, for the change in the direction of the apparent motion; for all that is necessary is to remember that the whole of the first forms of recession are with reference to the moving earth or its satellite—to whose motions they are directly referred; so that, as the terrestrial system moves through space, the sidereal heavens must pass round it, against its actual motion.

As soon as the motion of the sun, however, comes into question, these relations are at once changed, and another principle comes into operation—one very familiar to those observers who are in the habit of studying the consequences of motion. Any one who has travelled much by rail can hardly have failed to notice that, in inspecting the country as he passes rapidly by it, the most distant objects appear to advance with reference to any intermediate one—or, in other words, that the nearest object recedes from the more distant one with which its relative position is compared, as the observer passes it, being left behind by the passing train, with which the distant body seems to advance.

Now, this is precisely what happens whether in lunar recession or equinoctial precession,—that is to say (in the former), as the sun advances in space, the signs of the zodiac, or most distant visible objects, advance,—seem to advance with the advancing sun, and thus allow the terrestrial system, with its terrestrial observer, to fall slowly behind them: and hence it appears—

That the lunar nodes or terrestrial system recede upon the zodiac, because the earth is an intermediate object between the moving sun and relatively fixed zodiacal belt; so that it can be fairly held that in this, as in the previously noticed instances, recession is the measure of the period of a revolution,—that accomplished by the sun in passing round the focal body of its orbit.

The existence of a central sun has, of course, to be inferred, since that body has not, so far, been discovered; but there need be no difficulty here, for no body has, as yet, been recognised as revolving in an empty orbit, and it is contrary to all analogy that the sun should present the first instance of such a body.

Assuming the existence of a central sun, it would be

necessarily passing round the zodiac in the order of the signs, during the revolution of the sun, just as the sun passes round the zodiac in the same order during the revolution of the earth; this passage being a reflection of the actual passage of the sun round it: and in this passage it would carry the apsidal line with it.

Conceding these relations, and they amount to a philosophical demonstration, it becomes evident that the three associated phenomena of revolution of the apsides, recession of the nodes, and oscillation of the orbit, form a cycle of concurrent events,—a lunar cycle, whose period they determine; the interpreting significance of which is, that it is the measure of the time occupied by the sun in passing once round the central sun or focal body of its orbit.

But even beyond this the lunar apse can be made to speak, since the fact that two of its revolutions occupy a shorter period than one recession of the nodes is not without its special value; for—owing to the change in the direction of the apparent motion of recession, under which from passing from east to west, when the motions of the moon and earth are being considered, and thus occasioning the difference between sidereal and solar time, by rendering the former periods shorter than the latter, it passes from west to east when the motion of the sun (or other concentric body) is being examined—the sidereal heavens are now advancing upon the synodic revolution instead of receding from it, and hence the point of conjunction is regained before the sidereal relations are resumed. But through this significant fact it is learned that the body with which it is associated,—the central sun,—is itself in motion.

This, however, is not surprising; for it is clearly demonstrated by the terrestrial cycle (of precession of the equinoxes, revolution of the apsides, and variation in the obliquity of the ecliptic), which, from being in its three associated phenomena parallel to the lunar cycle, must have a similar interpreting value; so that not only the fact of the motion, but the actual period of the revolution, of the central sun, is defined.

Nor do the signs of motion drawn from apsidal revolution cease, even here; for the terrestrial apsides, in their period of some 21,000 years, indicate a recession in time from the period of precession (some 25,868 years). And since recession in time now stands for progression in space, or changing relations in the body to which it is imputed, it is thus shown that the body round which the central sun is passing is also in motion.

From these conclusions, which have been necessarily thrown together in a highly condensed form, it becomes evident that a careful examination of the phenomenon known as apsidal revolution can be made very fruitful in results; but these results only point to the outlying portions of a vast field of inquiry, extended research in which will more than amply repay the persevering investigator.

GEOMETRICAL DEMONSTRATIONS.

Fig. I.

The Centric Solar System passing across the Zodiac.

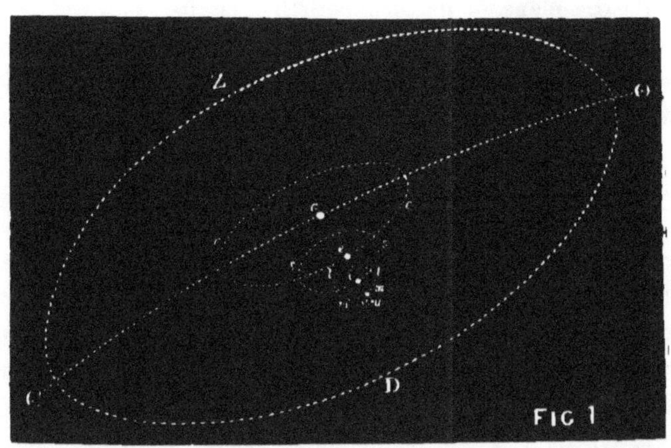

Z O D C. The zodiac.

C O. The path of the centric sun across the zodiac.

a. The centric sun.

e. The central sun.

i. The sun.

o. The earth.

u. The moon.

c c. The orbit of the central sun, round *a*, the centric sun.

s s. The orbit of the sun, round *e*, the central sun.

t t. The orbit of the earth, round *i*, the sun.

m m. The orbit of the moon, round *o*, the earth.

The centric sun, *a*, in moving from C to O passes *across* the plane of the zodiac, while the several members of its system are revolving on that plane, within the zodiacal belt.

It becomes evident from considering this diagram, that the plane of the zodiac represents the equator of the centric sun, and the paths of the members of its system, which are continuously passing round it within the circle of the zodiac, while the plane of its own path lies on its polar axis or at right angles to its equatorial plane; so that, while the members of its system are passing from west to east, it is moving in a direction which, with reference to the polar relations to space (absolutely), ultimately resolves itself into north and south.

The ultimate passage of the whole system from the zodiac, or the varying angular relations of the zodiac to the mean plane of the system passing through it, which must be occurring, unless compensating processes are at work, are not considered here. The time required to verify such changes would be incalculable, the data being as yet not ascertained, and the phenomena elicited coincident if not identical with those which establish the so-called motion of the sun, but actual motion of the centric sun.

Fig. II.

The varying relations of the Central Solar System to the Celestial Polar Centre, at definite positions on the Centric Sun's Orbit.

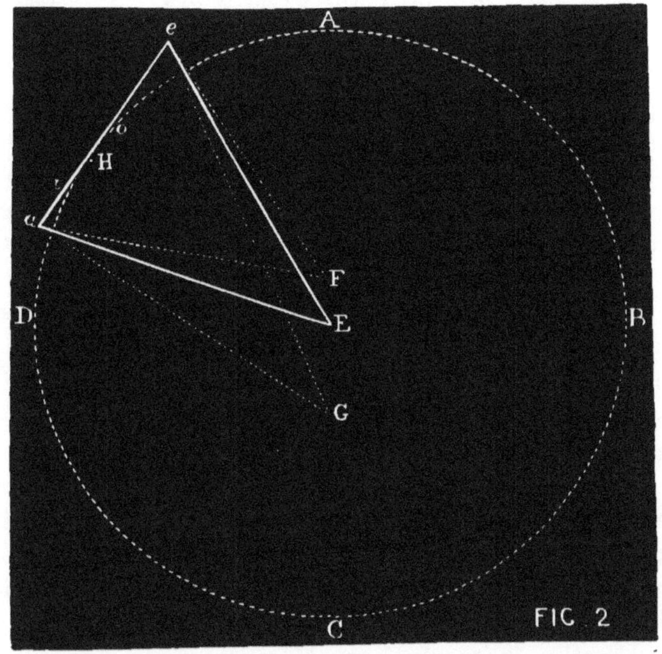

A B C D.	A circle representing the orbit of the centric sun.
E.	The centre of that circle.
F G.	Excentric or focal points within that circle.
H.	The centric sun.
a e.	The plane of the orbit of the central sun.
i o.	The equator of the centric sun.*

* The equator of the centric sun is parallel to that portion of the circle of its path at which it is found at any given time; hence, *e* being more distant from that circle than *a*, *a e* is necessarily inclined to *i o*.

a e. The central sun at opposite points of its orbit.

E *a*, E *e*. Radial lines showing the relative distance of *a* and *e* from E.

The line E *e* is longer than the line E *a*: therefore, at *a* the central solar system, and with it the earth, is nearer to E, the celestial polar centre, than at *e*. At G this difference increases, but at F it diminishes; that is to say, it increases with the increasing distance of H, and diminishes with its diminishing distance, from the focal body. A great principle is demonstrated by this variation in the difference, for, although at G the attracting force is more remote from *a* H *e* than at E or F, it acts with greater relative force on *a* than on *e*, and therefore tends to increase the inclination of *a e*; but at F, though nearer to *a* H *e*, it has increased its proximity, and with this the force of its attraction more rapidly upon *e* than upon *a*, and therefore tends to diminish the inclination of *a e*; so that the inclination of the orbit of the central sun is thus shown to diminish with the approximation of the centric sun to the celestial polar centre, and to increase with the recession of that body from it.

In virtue of this principle, the inclination of the orbit of the central sun will be least, when the centric sun is nearest to the celestial polar centre, and greatest when it is furthest from it; and also in virtue of it, when the centric sun is nearest to the celestial polar centre, there will be a maximum of elevation in the land in the northern hemisphere of the earth, or elongation of its polar axis; and a minimum oscillation in the level of the land as the central sun passes round the centric sun: whereas, when the centric sun is furthest from the celestial polar centre, there will be a minimum elevation of the circumpolar land, or elongation of the polar axis; and a maximum of oscillation in level during the central sun's revolution.

The variation in the obliquity of the ecliptic will be determined by the same law, but the oscillation in the plane

HALLIDIE LIBRARY OF THE UNIVERSITY OF CALIFORNIA

of the moon's orbit appears to depend upon the excentric attraction of the central sun. See *Orbital Motion*.

Another principle is taught by this diagram; for if the polar axes point to the centre of the celestial polar centre, and not to the centre of the orbit of the centric sun, then the equator of the centric sun would be inclined to its own orbit at all points of its course but those of greatest proximity to and remoteness from the celestial polar centre, this inclination varying in degree, progressively, with its position in space.

According to the law of projection established in *Eccentric and Centric Force* and *Orbital Motion*, the excentricity of the celestial polar centre in the orbit of the centric sun may be an indication that the path of the centric sun is the result of a single projection,—the measure of one act: and therefore the polar axes may all point to the true centre of the celestial polar centre, and cause the variation indicated in even all of the orbits; but these variations (if existent) are probably not commensurable, and must be distinguished from those recognised under observation, with which they would, of course, be blended.

In every case it is, perhaps, possible that focal excentricity may be a result of the law of projection; but if this were so, then each revolution would indicate a single subordinate act of projection, and each orbit would be traversed at a velocity which increased with recession from, and diminished with regression towards, the focus of motion.

FIG. III.

Apparent path of the Central Sun round the Zodiac, with the Moon's perigee accompanying, and always pointing to the Central Sun.

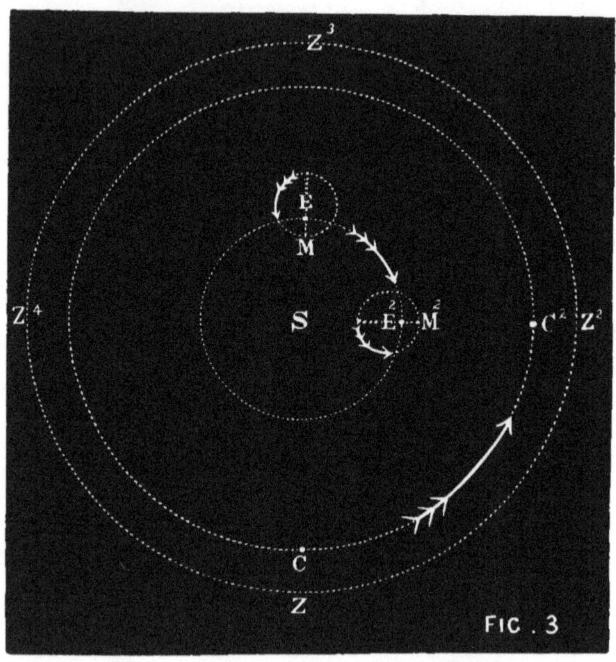

Z Z² Z³ Z⁴.	The zodiac or sphere of the fixed stars—the circle of comparison to which each revolution in space should be referred.
C C².	The apparent path of the central sun round the zodiac on the one hand, and the sun on the other.
S.	The sun, regarded as a relative centre of rest.
C C².	The central sun, at two of the quadratic points of its apparent orbit.

E E^2. The earth, at two of the quadratic points of its orbit.

M M^2. The moon in perigee, at two quadratic points of its orbit.

The terrestrial system being at E, with the moon, M, in perigee and conjunction, the sun is at S, and the central sun at C, all of them lying on the line E S C Z—Z being the point on the zodiac to which any change in their mutual relations will be referred; the moon, M, being between the earth, E, and the sun, S, while the central sun, C, is between the sun, S, and the zodiac, Z.

If the central sun, C, be now supposed to have traversed one quarter of its apparent path, it will be at C^2, between S and Z^2, its position in space being marked by Z^2. During its passage from C to C^2 it will have traversed one quarter of the zodiac, from Z to Z^2, carrying the moon's perigeal point with it, so that the moon's perigee will have advanced from M to M^2, also traversing one quarter of the zodiac, from Z to Z^2; while, during this double advance of the central sun and the moon's perigee, the terrestrial system will have receded from E to E^2, so that the moon's perigee is now in opposition, the earth lying between the moon and the sun: and thus the lunar apsides will have described half a revolution with reference to the sun, but only a quarter revolution in space.

Moreover, the points of the lunar nodes, when the terrestrial system was at E, were Z and Z^3, or, like the apsides, on a line with the sun and the central sun: they are now found to be Z^2, and Z^4, still on a line with the sun and the central sun, but in quadrature with Z and Z^3, having receded round one quarter of the zodiac, the one node from Z^3 to Z^2, the other from Z to Z^4.

Further, at E the sun lies between C, the central sun, and E, the terrestrial system, so that here the terrestrial system is in opposition to the central sun; but at E^2 the terrestrial system lies between S, the sun, and C, the central sun, or is in conjunction with the central sun: so that there is an inversion in the relations of conjunction and opposition, under which, when the moon's perigee is in conjunction with the sun, the terrestrial system is in opposition to the central sun; and when the moon's perigee is in opposition to the sun, the terrestrial system is in conjunction with the central sun.

FIG. IV.

The Revolution of the Lunar Apsides.

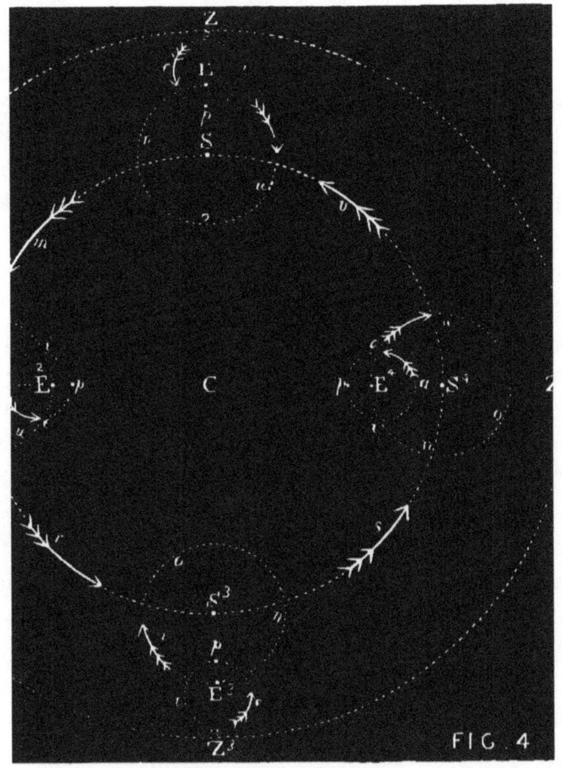

The zodiac, or circle of comparison.

The central sun, regarded as a relative centre of rest.

The sun at the quadratic points of its orbit.

E E^2 E^3 E^4.	The relative positions of the earth at the indicated positions of the sun in its orbit.
m r s v.	The orbit of the sun.
n u o.	The orbit of the earth.
a e p i.	The orbit of the moon.
a p.	The lunar apsides.

When the sun advances in its orbit, *m r s v*, from S to S^2 the terrestrial system retreats round the ecliptic, *n u o*, from E to E^2, the moon's perigeal point *p*, on *a* E *p*, simultaneously advancing round *a e p i*, the moon's orbit, to *p* on *a* E^2 *p*, because it always lies between the earth and the central sun.

At S the moon's perigee is in conjunction with the sun, because the sun now lies between the terrestrial system and the central sun; and the line Z Z^3 indicates the line of the lunar apsides and the position of the lunar nodes.

At S^2, on the other hand, the moon's perigee is in opposition to the sun, because the terrestrial system now lies between the sun and the central sun, the moon's perigee still being between the earth and the central sun; so that, although the sun has only traversed a quarter of its orbit, or passed from S to S^2, and along the zodiac from Z to Z^2, the perigeal point of the moon has passed from conjunction to opposition, or made half of its apparent revolution in advance: but when this apparent semi-revolution is measured on the zodiac, the moon's perigee is found only to have advanced from the line Z Z^3 to the line Z^2 Z^4, or from Z^3 to Z^4, between which point on the zodiac and the earth the moon's perigee now lies, so that it has in reality only advanced a quarter revolution in space, the other quarter revolution having resulted from the recession of the terrestrial system or lunar nodes, which now occupy the line Z^2 Z^4 instead of the line Z Z^3.

The apparent path of the central sun, were it visible or recognised (and the advance of the moon's perigee), is shown by the arc of the zodiac, Z^3 Z^4; for at E S the moon's perigee and the central sun lie between E and Z^3, but at E^2 S^2 they lie between E^2 and Z^4, having thus traversed one quarter of a circle in space as measured upon the zodiac, so that the actual path of the sun from S to S^2, or on the zodiac from Z to Z^2, must give an apparent path to the central sun or lunar perigee from Z^3 to Z^4.

The retreat of the node in the same period is shown on the arc Z Z^4, for at E S the terrestrial system lies between S and Z, but at E^2 S^2 it is between S^2 and Z^4; so that the node has made a quarter recession as measured on the zodiac, from Z to Z^4.

As the sun continues to advance from S^2 to S^3, the terrestrial system continues to recede from E^2 to E^3, while the moon's perigee continues to advance from *p* on E^2 to *p* on E^3, because it always points to the central sun; so that at E^3 S^3 the terrestrial system is once more in opposition with the central sun, the moon's perigee in conjunction with the sun: but this happens on the opposite side of the sun's orbit, for the sun has only passed through half of its path, the lunar nodes made only a half circle of recession, and the lunar apsides a half circle of advance; and yet the moon's perigee during this half circle of advance has passed from conjunction through opposition to conjunction again.

When the sun passes to S^4, the terrestrial system recedes to E^4, when it is again in conjunction with the central sun, —the lunar perigee in opposition to the sun, and at S the original relations are restored; so that while the sun has passed from S through S^2, S^3 and S^4 back again to S, the terrestrial system or lunar node has receded from Z through Z^4, Z^3 and Z^2 back again to Z, and the lunar apsides have advanced from Z^3 through Z^4, Z, and Z^2 back again to Z^3, or but a single revolution; and yet during this single revolution the moon's perigee will have passed twice from conjunction through opposition back again to conjunction, or made two apparent revolutions when measured from the sun, this apparent discrepancy having been occasioned by measuring the revolution of the lunar apsides from the sun instead of from the zodiac.

Were the central sun a visible body, then the terrestrial apsides would in the same manner describe two revolutions as computed from it, during the single revolution which they actually make round the zodiac.

OCEANIC TIDES.

PREFACE.

In the following pages an attempt has been made to explain the origin of the tides, upon simple and intelligible, and yet philosophic, principles. The view that the semi-diurnal tide is in reality caused by alternate equatorial and polar waves, originating simultaneously, and yet, from the circumstances of their genesis, following each other at some twelve hours' interval, approves itself to the mind as having reasonable grounds upon which to rest its claim for acceptance. The further view that the so-called diurnal wave, and the differences in the elevation of the respective semi-diurnal waves, originate in the superposition of direct and reflected waves—that irregularities or apparent divergencies from deductions drawn through any hitherto offered theory, whether in the equatorial regions or Pacific Ocean, arise from the relations of the respective coast lines to the direct waves, and the manner in which the direct, the reacting, and the reflected waves cross each other at definite points of their several courses—and that the absence of sensible tides is due to the want of an extensive concentrating or reflecting coast, is also a rational one, and deserving of consideration: while the conclusion that revulsion, due to the action of the individual gravity of the earth, and called into play by the inherent necessity for preserving its equilibrium as a spheroid of revolution, is a powerful agent in determining the synchronism of the regularly observed tidal phenomena, requires only to be pointed out for its importance to be recognised.

At the same time, in giving publicity to his opinions on these and other involved points, the Author is well aware that in some respects they embody generalizations drawn from insufficient data: but on such a subject this was only to be expected, and further investigations will serve to clear up all apparent obscurities and seeming inconsistencies. He does not profess to have devised a perfect theory: he rather claims to have traced principles through the instrumentality of which he hopes that such a theory will be ultimately attainable; but the demonstration of true principles, as guides to interpreting observed phenomena, is a very important step; and this he believes that it will be ultimately recognised that he has accomplished. He now submits the results of his inquiries to the criticism of his readers, that they may be fairly tested, and their true value determined.

HEREFORD, *March* 1865.

OCEANIC TIDES.

Great as has been the advance in scientific investigation of late years, and more especially great, as well as generally successful, the efforts that have been made to bring the results of these investigations, and the laws through which they are explained, to the level of the untrained mind, one branch of science at least can be pointed out which has not kept pace with the general onward movement; for, while on the one hand those most familiar with the subject admit that the ordinary scientific explanation reaches very few cases,* on the other a popular theory of the tides has yet to be written.

The general belief, in accordance with the view which is indeed still given in some of the best and latest practical astronomical works,† is that the disturbing body, as the moon or sun, attracts the water on that side of the earth which is the nearest to it more than it attracts the great

* See *Tides and Waves*, by G. B. Airy, M.A., F.R.S., Astronomer Royal, par. 14.

† See *A Handbook of Descriptive and Practical Astronomy*, by Geo. F. Chambers, F.R.G.S., p. 149.

mass of the earth, and therefore tends to raise the water from the earth on the side next to or immediately under it, but that it also attracts the great mass of the earth more than it attracts the water upon the side most distant from itself, and therefore tends to draw the earth from the water on the side most distant from itself: and thus produces exactly the same effect as if a force tended to draw the water away from the earth on that side. So that the disturbing body is held to raise the water on two opposite sides of the earth at the same time, and this by its direct action; the heaped up waters tending to a line passing through the centre of the earth and that of the disturbing body.

This theory is probably still offered to the unlearned public because of its simplicity, and because—although the greatest mathematicians and the most laborious observers of the present age have agreed equally in rejecting its foundations—they still compare all their observations with its results, in consequence of the singular agreement of these results with the actual phenomena, to explain which they are applied:* but, at any rate, from whatever motive, it is still put forward, and this is much to be regretted, because the prestige of authority is thus given to the dissemination of error; for an eminent living astronomer—the present Astronomer Royal—has said of this theory,† "That it is one of the most contemptible theories that was ever applied to explain a collection of important physical facts. It is entirely false in its principles, and entirely inapplicable in its results."

This theory has been called the Equilibrium theory, and it is known by that name, because its results are drawn from the assumed conditions of water at rest.

Laplace, in his investigations utterly disregarding this

* See *Tides and Waves*, par. 64.

† See *Ibid.*

theory, took a different starting-point, and very properly and sagaciously treated the tides as fluid in motion; but even his investigations, though they are allowed to be one of the most splendid works of the greatest mathematician of the past age, offer for the most part results rather of a negative than of a positive kind; for he, at last, took refuge in the assumption that all that we are certain of is, that the disturbances of the seas will be periodical as the forces that cause those disturbances, but that their times of maximum or minimum are not necessarily the same as the times of maximum or minimum of the forces,* which any accurate observer could have told him before he commenced his investigations.

The present Astronomer Royal, in his elaborate treatise on *Tides and Waves*, advances a wave theory in explanation of tidal phenomena: but though, in speaking of it, he says that he hopes it will be found that something has been added to the preceding investigations of motion, possessing in some degree a practical character, still he admits that, even in the state in which he leaves it, it reaches very few cases;† and the impartial examiner of his investigations is reluctantly compelled to admit that this is the case, though it has been said of his theory, that it has entirely superseded every previous theory on the subject, and that it will prove for many years to come the only sound foundation of our knowledge of the theory of the tides.‡

In the course of his investigations, he calls attention to an interesting fact,—which, however, might have been predicated from *primâ facie* evidence,—that in a tidal river affected by a general current towards the sea, and subject to friction, the low water in the upper part may be higher than the high

* See *Tides and Waves*, par. 121.

† See *Ibid.*, par. 14.

‡ See Advertisement to the *Encyclopædia of Astronomy.*

water near the sea.* This fact is a very important one, as bearing upon the whole subject, for it proves that river tides are derived tides, or true waves, having special laws of their own, which very possibly differ from the laws governing the motions of the great primary or oceanic tides; so that a theory, drawn from the study of the wave current *in canals*, and thus specially applicable to this form of derived tidal wave, may be inapplicable to, and therefore unfitted for, the interpretation of the laws governing the oceanic tides.

It is clear, therefore, from the admission of the most recent, as well as confessedly most eminent, authority on the subject, that an adequate scientific theory of oceanic tides is still wanting, and, as a consequence of this, that a popular explanation of them has yet to be written.

The ordinary tidal phenomena—such as the manner in which the most easily observed or semi-diurnal tide, in the progressive variations in its periods and range, follows the varying relative positions and inclinations of the moon and the sun—are so well known that they need not be dwelt upon here; for the intimate relations between this tide and the phases of the moon clearly establish the fact that, regarding the tides as evidences of disturbed equilibrium in the earth's fluid particles, the moon and the sun, in their changing relations with the surface of the earth during its axial rotation, are the disturbing bodies. They do more than this, indeed; for, since in the semi-diurnal tides the influence of the sun only makes itself felt by diminishing the range of the tide during the moon's passage to quadrature, and increasing it again during its return to the syzygies, in a manner that is well understood, the lunar tide is thus learnt to be the only one that, in the north of Europe at any rate, is commonly sensible: and thus it is perceived that the disturbing influence, whatever may be its nature, chiefly emanates from the moon.

* See *Tides and Waves*, pars. 343 and 508.

Under the theory of gravitation, the influence exercised by the disturbing bodies is attributed to the attracting power dependent on, and inherent in, their mass; and although this view has been disputed by some, even in the present day, it is found to account for and satisfactorily explain a larger series of observed tidal phenomena than any theory that has been proposed in its stead, and therefore will be taken as the basis of the following inductions.

The preponderating influence of the moon as a disturbing body, is, by the theory of attraction, at once accounted for, since the law of gravity being that its attraction diminishes inversely with the squares of the increasing distance, although the mass of the sun is so much greater than that of the moon, the latter is so much nearer to the earth than the former, that *through proximity* its disturbing force is rendered the more potent of the two.

This is the first induction drawn from the theory of gravity. In its application it will presently be found most important, not only in determining the relative strengths of the moon and the sun in their action upon the water of the earth, but actually in limiting the sphere of the action of the moon; for, it being granted that the nearer the attracting body the greater its attracting influence, it follows that the attracting influence of the earth, referred to its centre, upon the water on its surface, lying between it and the moon, and therefore exposed equally to the direct influence of both, through acting from only $\frac{1}{60}$th of the distance through which the moon exercises its power, must be proportionably greater than that of the moon; so that, when it is remembered that, in addition to this, the mass of the earth vastly preponderates over that of the moon,* it is at once perceived,

* The mass of the earth being regarded as unity, that of the moon is computed as 0·0114, so that it would take very nearly 88 moons to make one earth, their densities being equal.

and must be unhesitatingly admitted, that the moon cannot heap up the waters of the earth under itself,—that is, on a line between its own centre and the centre of the earth; since, when two similar forces are acting in a similar manner, but in direct antagonism to each other, the action of the weaker force must be controlled by, and subordinated to, that of the stronger; so that a single demonstration, drawn from an extension of the very principle upon which it rests to the phenomena which it has been supposed to explain, at once proves the fallacy of the Equilibrium theory, based as it is upon the view that the water is at rest.

This being the case—and that it is the case can hardly be denied—the question now very naturally suggests itself: Since the attraction of the earth is so much stronger upon anything lying on its surface than that of the moon, both through preponderance of mass and proximity, how can the attraction of the moon disturb the equilibrium of the water spread over the surface of the earth?

Difficult as the problem involved in this question may at first seem, it is in reality susceptible of a very ready and, at the same time, intelligible solution, for all that is necessary is to remember that it is the *direct* antagonism of the attractions of the earth and the moon that neutralizes the latter; and since the influence of gravity in each instance radiates as from a centre, this direct antagonism *can only exist on the direct line between the centres of the moon and the earth,** diminishing rapidly from this to the circumference of the hemisphere of which this right line of direct antagonism may be taken as the centre;† so that *the angular relations of any given point of the surface of this hemisphere to the relative centres of the earth and the moon* are very important elements

* That is, at B in the figure.

† That is, in passing from B towards A or C in the figure.

in determining the relative strength of the disturbing force that can, at any given time, be exercised by the latter.

Through this induction the very significant result is gained, that the attraction of the moon, or of any other similarly disturbing body, does not tend to draw the water, spread over the surface of the earth, from its centre, but imparts motion to it, introducing a circulation which commences at the periphery of the disturbed hemisphere, and is directed towards its superficial centre, thus primarily affecting its surface. Hence the lesson is learnt that every attracting disturbing body acts by introducing motion amongst the disturbed fluid particles;* the general tendency of which motion, when unmodified, is, in the first instance, towards the superficial centre of the disturbed hemisphere.

This view at once introduces an important modification in the otherwise universal law of gravity, for it is now seen that when acting through free space from a remote mass upon the surface of a spherical or spheroidal body whose inherent attraction is opposed to its own on the direct or common line of their mutual attractions, the attraction radiating from the remote mass, which unopposed would diminish with the increase of distance, through the introduction of the antagonising similar force against which it is now acting, actually inverts this law, since, as its radiating lines of force diverge from the common line of direct antagonism, their disturbing action on the surface of the spherical or spheroidal body increases with their increasing distance from the centre of action.

* So that it does not attract *vertically* from the centre of the disturbed body, but *obliquely*, in a direction more or less inclined to the surface at the point of attraction, the motive power increasing with the increasing obliquity till the right angle is reached, where a maximum is attained.

This very important principle can be easily illustrated and established geometrically; for if A B C be considered to represent the outline of the surface of a disturbed hemisphere, E its centre, B its superficial centre, and D the disturbing body, then the right line D E, passing between the centres of the disturbed and the disturbing bodies, will represent the line of common action or direct antagonism, along which, at B, the disturbing influence of D will be *nil;* from which point, as the disturbing radius D B increases progressively in length to D A or D C, although the actual quantity of attracting force from D is as progressively diminishing in a geometrical ratio to the points A C, on the lines D A, D C, at which points it attains its relative minimum, yet, owing to the ratio of angular inclination between the lines E A and D A, the relations of antagonism, which have been as progressively diminishing from E B to E A, at A are *nil;* so that the actually weakened force D A has, from being unopposed by the force E at A, attained there its greatest power of expressing itself, and hence acts at A most readily; through its action, imparting motion along the superficies A B towards B, to any fluid particles subjected to its influence there.

It follows from this that the disturbing attracting body acts upon the fluid particles of the disturbed body, not by drawing them further (or lifting them) from its centre, but by imparting motion to them round, or, as it might not inaptly be termed, parallel to, that centre, in a direction towards the central line of common action; so that the primary source of the tides is motive, *not lifting*, in its expression, owing to the circumstances under which it is brought into operation.*

A singular philosophical inference flows from this, no less than that if the earth and the moon could be regarded as fixed bodies in space,—that is, in themselves absolutely motionless,—the attraction of the moon would, under these circumstances, introduce motion into the waters of the earth, by drawing them, or causing them to flow, towards that point of the surface of the earth, or rather of the ocean, that was immediately under itself.†

This purely ideal view of the subject will be found to be of great value in further investigations, for it practically illustrates the origin of the lunar and other similarly caused tides; but though the flow thus produced would, under the imaginary fixed relations that have been indicated, be persistent, it will be convenient to regard it as expressing itself under the form of a succession of waves which, commencing from the periphery (circumference) of the disturbed hemisphere, would progressively advance as a series of concentric circles, increasing in magnitude (within certain limits) as they converged, as well as in the rapidity of their flow; so that the direction of their flow would be correctly indicated by a series of radiating lines converging in an equal degree upon a central (superficial) point.

It would appear, at first sight, as though this view were

* That is, the disturbing force is not—cannot be—vertical in its action, but *oblique*, according to the principle that has been laid down.

† But as the attraction of the earth would prevent the advancing waters from accumulating or heaping themselves up under the moon, a reflux would be originated in the form of under-currents, to preserve the equilibrium of the earth, so that the double attraction would thus introduce a double or to-and-fro motion—an actual circulation of the waters of the earth.

directly opposed to the principle which has been just laid down, that the attraction of the moon cannot heap up water under itself, since the uncorrected inference would be, that the progressive increase in the magnitude of the waves, as they converged, must necessarily lead to the piling up, or accumulation, of the waters directly under the moon. This, however, has been shown to be impossible: hence, even at this early stage a modification has to be introduced into this primary simple theory; one consistent with, and dependent upon, the antagonising causes which are at once originating and controlling the imparted motion. This is to be found in the principle that this progressive increase in magnitude cannot go on beyond a certain point,* which in this supposed fixed, disturbed hemisphere would be indicated by a circle drawn some 45° or thereabouts from its periphery. Hence, in this disturbed hemisphere the concentrating waves would within certain limits progressively increase in magnitude to the 45°, at or about which point they would reach their maximum.

The reason why this should be the case will at once appear, on considering the properties of the assumed sphere, whose surface is supposed to be acted upon by the attraction of a body influencing it from without; for this attraction, in obedience to the law of gravity, will increase rapidly in strength with the rapidly diminishing distance up to 45°; but very slightly and slowly from that point, owing to the small amount and gradual nature of the diminution in distance that now takes place; whereas, the antagonism of the attraction of the disturbed body referred to its centre, which is equal to zero at the periphery, or circumference of the superficies of the disturbed hemisphere, where it is all but

* Because the centric force of the earth will cause the accumulating waters to subside and flow off in the form of under-currents, whose direction will be towards the point from which the superficial wave started, to supply for the water which rises there, in order to preserve the general equilibrium.

inoperative, will increase very slowly up to 45°, owing to the slow variation in the angular relations of the two forces, which are still considerable, but from that point it rapidly increases its power of expressing its action, with the rapidly diminishing angle, to its maximum of entire control at the central point, where the angular relations cease in the right line of direct antagonism.

The consequences of these relations are now intelligible enough; for while the direct action of the disturbing body, so far almost unopposed, causes the successive tidal waves to flow towards the 45th parallel, where the highest (direct) tide will be gained, the hitherto all but inoperative, but now less perfectly, and gradually less and less opposed, centric action of the earth coming into play diffuses the accumulated waters again, and so prevents the culmination of the tidal wave reaching the central point of the surface of the supposed immoveable hemisphere.*

Under these imaginary relations the tendencies of the two forces are brought out into bold relief, in order to show that the unmodified action of the disturbing body is to heap up, and the centric action of the disturbed body to diffuse, the tidal wave; but, under the view that has been, so far, examined, the actual phenomena, as observed in nature, have been inverted, the periphery of the disturbed hemisphere having been regarded as diffused in its persistent relations, the centre of course as concentrated,—and this it is which has created the difficulty in applying the principles that have been indicated to the actual facts which they will presently have to explain.

The manner in which these artificially constructed difficulties disappear, in the actual phenomena that have to be

* The attraction of the earth at all times tends to diffuse those waves which disturbing action would otherwise accumulate. Hence, direct tidal waves will, by diffusion, be scarcely appreciable in their mensurable range; a combination of other causes, and especially reflection, being necessary to render them sensible as tides.

interpreted, is very remarkable; for, impart a rotating motion to the hitherto fixed sphere, or let the earth, as it does in fact, revolve upon its axis, and the diffused periphery of the disturbed hemisphere, hitherto represented by a great circle of the sphere, is at once concentrated into two points, the extremities of the axis of rotation or poles of the earth,* while the hitherto central point, towards which the disturbing body acts, is itself converted into a great circle of the sphere.† This is a remarkable change to have been produced by the simple act of axial revolution, but its consequences will be found yet more remarkable than they probably at first appear.

They must be examined through the known tendencies of the two forces that have been now seen to be in operation. It has been shown that the disturbing body exercises its direct action by introducing motion (or causing a flow of the tide) from the periphery (or circumference) of the hitherto considered fixed hemisphere towards its centre. It has also been shown that the introduction of motion, round an axis of the hitherto stationary sphere, transfers the periphery of the fixed hemisphere, with all the persistent relations and tendencies which attach themselves to it, to the two extremities of the axis of revolution.

The first consequence of this is a very remarkable one, for the action of the moon upon the superficial waters of the earth having been proved to be to impart motion to them in a determinate direction, it is now clear that this direction must primarily be from the poles towards the equator; while the centric force of the earth having been proved to have a tendency to diffuse the accumulating waters over a surface

* The persistent local relations to the disturbing body are, of course, here referred to.

† That is, the equatorial regions, so that the introduction of axial rotation not only does away with the difficulties involved in the concentrating influence of a disturbing body, but actually diffuses the effects of disturbing attraction.

which may, for facility of description, be termed the equatorial regions, and forms an extensive belt encircling the earth, it follows, first from the relations of areas, and second from the relations of the 45th parallel, that the motion and range of the tides, when unmodified, will be most sensible in the circumpolar regions, and become necessarily insignificant in the equatorial.*

This diffusing power of the centric action of the earth deserves careful study. It is not immediately destructive to the motion of the waters advancing from the polar regions, but spreads out, or progressively widens, the area of the motion. Hence, in the first instance, the wave initiated in the polar regions will have a radiating character as it approaches the equator; but the radiation, as will be seen presently, will tend westwards.†

In consequence of this property of the centric action of the earth, it can and does offer no direct barrier to the advancing waters when they reach the equator; hence, the tidal wave coming direct from the polar regions and progressively increasing its motion, at least to the 45th parallel, owing to the progressive increase in the strength of the force under which it acts, will, in obedience to the laws of projection and motion, although it reaches the equator as a much thinner wave, not die out there, but pass on into the opposite hemisphere ‡

From this tendency the first distinction in the character of the tidal wave is learned; for it is now found to be *direct* from the polar regions towards the equatorial, and *reactive* from the equatorial regions towards the poles.

* This only applies to the direct tidal waves, and does not affect the actual or observed tides, except in so far as it shows the tendencies of particular regions of the revolving sphere.

† This tendency will probably hardly show itself till the equatorial regions are approached, since in the temperate regions the tendency in the atmospheric and oceanic currents is eastwards.

‡ That is, unless obstructed in its course.

It has been shown that the introduction of rotation to the earth concentrates the continuous direct disturbing action of, say the moon, in the polar regions.* This is only strictly accurate when the poles are regarded as points of persistent direct action (which they alone are), for the whole of the hemisphere under the moon is really subject, more or less, to its disturbing influence; but rotation modifies that influence in this wise: the meridian for the time being under the moon may be considered to divide the moving disturbed hemisphere into two halves, one of which, the eastern, *quoad* its angular motion, is opposing and passing away from its direct action — moving against it; while the other is moving with, or in the same line as that direct action expresses itself on the surface of the earth. The consequence of this is, that the rotation of the earth neutralizes, or renders insensible, the direct disturbing action of the moon (*quoad* wave genesis) westward of the meridian, over which, for the time, it is to be found, and intensifies the sensible expression of that action eastwards of the same meridian.†

From these remarkable relations, another important phenomenon in the history of the genesis of the tidal wave is found,—that, in the equatorial regions, the direct action of the moon causes a wave of water to follow it in its apparent course, or pass from the east in a westerly direction.‡ This wave, though really direct, *quoad* the action of the moon, can, from its relations to the axial motion of the earth, be very properly termed the *receding* wave.

* That is to say, practically reduces its persistent direct power of expressing itself to a minimum of surface; in reality to two points,—the poles.

† This is self-evident and universally admitted. See *Tides and Waves.*

‡ This wave, judging from its relations to the new and full moon on the eastern equatorial regions of the coasts of Africa and South America, follows the moon at an interval of 4 hours—or 60° in distance—which represents the angle at which the moon exercises a maximum of disturbing force.

The centric force of the earth will tone this wave down, as it does all regular tidal waves in the equatorial regions, causing it to diffuse itself towards the north and south as it passes westwards; thus giving it, as a whole, a radiating tendency,—those portions which pass off towards the north or the south forming, in reality, reactive waves, similar in character to those already described, though differing in the direction of their origin.*

The two extremes of the primary modification introduced by the axial motion of the earth into the tidal wave have now been shown. Their leading features deserve careful study, more particularly that, in their directions, they are transverse to each other; the polar waves primarily flowing with the meridians, or north and south; the equatorial passing with the parallels of latitude, or in the line of the equator towards the west. But it must be remembered that, from these extremes, as the points of genesis of the wave approach each other (in passing respectively north and south), the direction of motion gradually changes, until, at the mean, it has become parallel each to each, or merged in the same line; so that (theoretically), regarding the surface of the waters from the moon, the general direction of the lines of motion would be radiating from a point under the moon to the eastern circumference of the visible hemisphere, straight at the equator and the polar extremities,† but gradually curving from these towards the mean, with the concavity of the curves on either side of the equator directed towards their respective poles.‡

* That is to say, these originate in a wave moving parallel to the equator; whereas the others proceed from waves moving towards the equator.

† Straight at the equator, because the motion there is always in the line of attraction—straight at the poles, because there the change in direction is the slowest—curved at the mean, or 45th parallel, because from the 60th to that parallel the oblique action of the moon is most felt, and therefore the change in direction from the polar meridian to the equatorial parallel most rapid.

‡ Because at the poles the moon is always acting in the direction of the

In noticing the transverse relations of the extreme directions of the tendencies of motion of the direct tidal wave at the equator and the poles, it is worthy of remark, that the results reached here are the exact reverse of those previously drawn from theory; for Professor Airy, in examining Laplace's investigations, says (*Tides and Waves*, par. 103):—

"Thus we obtain the result, that at the equator the water moves only north and south, resting for an instant at the change of motion: in every other part of the earth the water is always moving with some velocity; but the current is perpetually changing its direction. At the poles the velocity is constant, and the direction is always transverse to the meridian which passes through the luminary."

A very little reflection will show that this latter view cannot be correct; for the direction of the general tendency of the motion must be determined by the length of time that the line of direction of the originating force is actually or comparatively undisturbed. Now, at the equator this force always acts transverse to the meridian, so that the direct tendency there cannot be north and south; while at the poles the velocity of rotation is so low, as compared to that at the equator, that the direction of the originating force is changed very slowly, so that the tendency of the general direction of the motion (during a short interval of time) will much more nearly approach that of the meridians than one transverse to them.*

meridian over which it is crossing at the time; but as the polar waves pass towards the equator, or north and south, the moon in moving to the west draws them after it, seeing that they are becoming relatively to the moon more and more east of it.

* Instead of following the moon round the pole, as the mathematical theory infers, the wave will advance to the equator, and be emancipated from the power of the moon's attraction as the moon passes the meridian 90° from its meridian of genesis (or even sooner); hence, from the proximity of the meridians to each other in the polar regions, the scope for curve is very slight in proportion to that for extension.

In considering the general tendency to a mean curvilinear character, in the direction imparted to the moving waters by the axial motion of the earth, another consequence has been incidentally brought out, which must be now more distinctly referred to. The extreme direction of this tendency has been shown to be at the poles, primarily north and south, and at the equator invariably westward. Hence the impulse imparted at the poles is primarily north and south; but this impulse is being constantly (and at first even more vigorously*) renewed in a modified direction, so that, as the wave of water passes towards the equator, it is warped towards the west; this warping—if it may be so termed—being caused by the axial motion, which is constantly changing the direction of the renewed impulse. The consequence of this is, the extremes of direct action being at the equator and the poles, that at the 45th parallel, or thereabouts, a mean is gained; the unmodified direction there (or in the circumpolar regions of which that parallel may be held to represent the centre) being S.W. and N.W. respectively. From this mean point, proceeding towards the poles, the polar direction will be progressively increasing, and towards the equator the equatorial.

Now, this 45th parallel will not only represent a mean point, but a mean area, forming, as it does, the centre of the circumpolar regions.† Hence, at this parallel, three significant relations are combined—a mean of area, a mean in direction, and a maximum in height.‡

* As regards the ratio of increasing strength of the attraction of the disturbing body.

† The temperate regions are called circumpolar, because of their relations to the polar regions. A designation drawn from geometrical position is perhaps more convenient than one taken from physical relations.

‡ This maximum in height refers to the height produced by disturbing attraction: it extends from the 60th to the 45th parallel; although a maximum relation, it will hardly produce a sensible effect on the observed (or reflected) tides, unless by combining or coinciding with them in some regions or localities.

Read through these relations, the first inference is, that in the circumpolar regions the direct tidal wave will attain its greatest elevation or range; that in the northern hemisphere, its mean, unmodified direction will be towards the S.W., and in the southern towards the N.W.; the westerly direction being, moreover, increased in each instance by the tendency which the waves will have, as they advance towards the equator, to lag behind the rotating surface of the earth.*

Hence, for convenience of interpretation, and yet with great general accuracy, the mean or circumpolar may be taken as the direction of the direct tidal waves passing from the polar regions towards the equatorial; while in the reacting tidal waves, which pass from the equator towards the poles, this transmitted direction will be continued—that is to say, in the northern hemisphere the reacting waves will pass to the N.W., and in the southern to the S.W., or exactly cross the direct waves; † but here, since the waves are passing from a high to a lower rotating velocity (as regards the earth's axial motion), they will have a tendency to advance upon the rotating earth, or incline eastwards of their direct course.‡

A clear view of the mean direction of the primary or direct tidal wave can now be gathered, according to the

* In the circumpolar regions, or temperate zones, the tendency is, in reality, to advance upon the rotating motion or pass eastwards, as is seen in the atmospheric and oceanic currents in these regions. Hence the theoretical tendency to *lag* behind the earth's axial motion cannot come into play until the equatorial regions are approached, where true recession will commence. In waves passing from the equator to the poles the tendency to move eastwards will be increased as the wave gradually passes from a higher to a lower rotatory velocity, by its carrying a proportion of the higher velocity with it.

† It is well known by all observers, that waves cross and re-cross each other in all directions and at all angles, without interfering with each other's course.

‡ So that, in the circumpolar regions, they will increase the general tendency to pass eastwards.

parallel through which it is being traced, since it is found to resolve itself into three principal waves, each possessing distinctive characteristics drawn from the position of the region of its genesis.

1st. *The northern direct* — which passes through the northern circumpolar to the equatorial regions in a S.W. direction, and, crossing the equator, enters the southern hemisphere as a reacting wave;

2nd. *The southern direct* — which passes through the southern circumpolar to the equatorial regions in a N.W. direction, and, crossing the equator, continues its course through the northern hemisphere in the same general line, though slightly diminishing its inclination westwards as it advances to the north; and

3rd. *The equatorial direct, or receding* — which is found in the equatorial regions, and passes to the west in a direction generally parallel to the equator; which also sends off reacting waves during its passage, which are radiating in their origin, but more or less inclined, in the northern hemisphere, to the N.W., and in the southern to the S.W.*

It must be remembered that each of these waves, when taken singly, and in its unmodified form, has, in reality, a comparatively slight range; † so that even collectively, as will be seen presently, they can be only the primary cause of the observed tides; the nature of the actual tides is to be learnt from the manner in which these primary waves are modified by combination, and otherwise. Hence the modifying causes of the primary waves have now to be considered.‡

* So that the course of the direct tidal wave will be definite for every parallel, and can be determined in any region of the globe by the parallel which it is crossing and the range through which it has already moved.

† In elevation.

‡ Although the *direct* tides are treated as waves, they are perhaps more correctly periodic currents flowing through the oceans in the directions indicated, and only seem to make themselves sensible as tides by flowing on to an extensive opposed coast line after a long range.

It must be conceded that the primary waves can only be expected to be traced in the ocean, or rather on their flowing on to an oceanic coast, after passing over an extensive and uninterrupted surface of water; and that, in reality, even here, they have not a sufficient range to make themselves, in all cases, readily sensible in an unmodified form;* indeed, in their purity, they could only be present if the revolving earth were covered with a tolerably uniform and deep layer of water; under which conditions the ordinary means of detecting them would be wanting.† Hence, theoretically, they can only be used to indicate the direction towards which, at a given point of the earth, the unmodified wave would tend, in order thus to learn how modifying causes can be, and most probably are, brought into play; but thus used, they will be found to be of very great value.

It is at once evident that the principal modifying cause is to be found in the configuration of the several continents which form the boundaries of the oceans and seas into which the collected waters of the earth are, for the most part, divided; which, by opposing obstruction to the free passage of the waves, change the direction of their course, *reflecting* them, and thus producing a change of motion in another, also determinate, direction. Hence in *reflection*—a form of recoil from an obstructing boundary—the true source of a large proportion of the observed tides, and certainly of the observed semi-diurnal tides on the coast of Europe, will be found.

The general law of reflection—that its angle will equal the angle of incidence—is well known, and need not be

* Thus in Europe, unless they aid in forming the tides in the North Sea and on the west coasts of Scotland and Ireland, they have not been observed; although in the equatorial regions of the east coasts of America and Africa they form the observed tides.

† Because, under these circumstances, there would be no coasts for them to flow on to or past.

dwelt upon here. Its unmodified expression in the ocean would be this, that if the equatorial plane of the earth could be converted into a solid obstacle to the passage of the waves of water, then in the northern hemisphere that flowing from the N.E. would be reflected to the N.W., and, in the southern hemisphere, that from the S.E. to the S.W.; while, if a solid meridional plane could be passed through its polar axis, then the receding wave coming from the E. would be reflected back upon itself, or towards the E., while that coming from the N.E. would be reflected towards the S.E., and that from the S.E. towards the N.E., each variation in angle making itself felt.

Simple as these laws are, however, in practice, it is very difficult to apply them, owing to the generally irregular form of the several boundaries of the ocean, and owing to modifications in them, caused by friction, on the one hand, and the tendency, determined by the axial motion of the earth, to introduce certain definite currents or lines of direction, on the other. Other causes also will, necessarily, be in operation, such as those which govern the laws of projection under which the waves of water are moving, and determine the application of the secondary forces engaged in producing them; but any reference to these would be widening the field of discussion too much. They must, therefore, be passed over as subordinate to the great primary causes which are at work; but, perhaps, the most important is to be found in the fact that the reflected water is reflected in a medium of equal density to itself—that is, in itself—for the reflection takes place in water; so that when it occurs at an angle, the line of reflection will be more or less curvilinear. Hence the reflection of the tidal wave becomes a special, and even complicated, study; but, notwithstanding the difficulty of the analysis of the problem involved in this branch of the subject, certain guides can be found of great value: thus, in the circulation of the atmosphere, it is now a well-known empirical fact that, while in the tropics the prevailing current is from E. to W., owing to the tendency in the fluid particles to lag behind the diurnal motion, in the temperate zones this order is inverted, the current here moving from W. to E., or advancing on the diurnal motion. This, doubtless, is caused by the earth's axial motion, and most probably depends, at least in part, upon the different actual velocity of rotation at the respective regions indicated; but, if this is the case—and in the present state of knowledge on the subject it can hardly be called in question—then the same tendency will be present in the circulating waters of the ocean, and though, owing to the restraining action of the several coast lines, it cannot express itself in the same manner, nevertheless it will be acting as an important modifying cause in the determination of the great oceanic tidal waves.

In the actual ocean currents, therefore (though these are quite distinct from the tides, and are the expression of the uniform, as the latter are of the varying, circulation of the earth), guides to the true direction of the tidal wave will be found; thus, take for an example the Atlantic Ocean, which, from being the best known, furnishes the most readily intelligible illustration of the principle which it is now sought to establish. In this ocean the equatorial current flows from E. to W. On reaching the coast of South America it finds, in that coast, a more or less triangular or wedge-shaped obstacle, which divides it into two streams, one of which passes towards the Gulf of Mexico, the other towards Cape Horn. The apex of this obstacle, however, is not on the line of the equator, but south of it some 6° or 7°. Hence the great mass of the equatorial current will necessarily be directed into the Northern Ocean. This is the stream which has now to be considered. Guided by the general outline of the coast, rather than reflected from it (the direct reflection being impeded by the pressure of the water outside that which reaches the coast line), it proceeds in a general north-westerly direction past the West Indian Islands (sending offshoots, or derived currents, on its course into the Caribbean Sea and Gulf of Mexico) to the coast of North America, directed and guided by which, again, and aided by the circulating tendency (eastwards) of the region which it is gaining, it now moves to the N.E., and, in the form of the Gulf Stream, reaches the coast of Europe, where again, divided by the British Isles, it sends its streams north and south.*

Under this view, the accuracy of which is hardly questionable, the Atlantic equatorial current and Gulf Stream become a single continuous river in the Ocean, sending out branches and changing its direction according to intelligible laws. But, if this is the case, then the oceanic tides, which only differ from the oceanic currents in being intermitting in their flow, will obey the same laws and tend towards the same direction: thus, supposing the moon to be full on the meridian of Christiania, its action will be vertical in the equatorial regions on the western coast of Africa, so that its disturbing power here will be *nil*, though in the polar regions it will be in full operation—hence the receding equatorial wave has not yet commenced, though the direct polar waves are flowing; as the equatorial regions pass to the east under the moon,† the receding equatorial wave will commence its flow, passing to the west,

* So that the circumstances in which it is called into play on the Atlantic coasts convert the reflecting influence of those coasts into guiding, rather than direct reflecting, media.

† Many theorists confuse the motion of the moon with its apparent motion. Its actual motion is from west to east, and is the cause of the diurnal variation in the time of the high water of the semi-diurnal tides. Its apparent motion is from east to west, and is a reflection of the axial motion of the earth. Hence the axial motion of the earth is the cause of the crest of the tidal wave lying always to the east of the moon, on the equator, and passing from east to west.

towards the meridian over which the moon is lying, the mean of which will probably be some 45° from that meridian; the polar waves in the meantime passing north and south respectively, inclining to the W., and then from the circumpolar regions passing respectively N.W. and S.W. Disregarding the origin of these waves, and now looking upon them as simply three waves moving at the same time, the one due W. on the equator, the other N.W. in the South Atlantic, and S.W. in the north —What will their several courses be?

Studying these through their several directions on the terrestrial globe,* they will be easily followed; for, first the receding wave, like the great equatorial current, will, on impinging on the wedge-shaped coast of South America, divide into two waves, by far the larger of which will pass into the northern hemisphere, in the track of the Gulf Stream; then the South Atlantic wave (owing to the general inclination of the passage between the two oceans being in the direction of its own course) will pass, almost entirely, as a receding wave, into the North Atlantic, following the equatorial wave; while the North Atlantic wave, sweeping in a south-westerly curve across the Atlantic, inclining more and more to the west as it approaches the equatorial regions, will be reflected, almost entirely, on the opposite curve back into the northern hemisphere, following the general direction of the equatorial wave, and therefore accompanying the southern reacting wave.

It would thus appear that in the Atlantic Ocean these three waves pass from the equatorial regions into the North Atlantic, where, after flowing round the coast of America and carrying the flood-tide with them, they cross

* Diagrams are not used, because, from their flat surface, they can be made to give but a very faint idea of the true relations. A terrestrial globe should be referred to by the reader.

the Atlantic, in the direction of the Gulf Stream, and so bring the flood-tide to Europe.*

If this is the case, then the oceanic co-tidal lines are at once shown to be inaccurate, even in their name, since the same wave cannot reach the west and east coast of the temperate zone of the North American Ocean at the same time. They rather point to the order of succession, and help to measure the time of transit of the tide from the one coast to the other.†

The general course of the three waves having been thus briefly indicated, their relations to each other have now to be examined, in order, if possible, through these to learn the origin of the North Atlantic semi-diurnal tides; since their combination, under the views now given, would only account for a single tide in each sidereal day. These relations are to be learnt by considering the relative spaces (linear areas) that have to be passed through by each from their point of genesis to their common point of reflection,—determining the period at which the shortest and most direct wave reaches this common point,—learning its velocity of motion by computing how long after the transit of the moon over the meridian of this common point it reaches it, and so discovering, approximately, how long it has taken to perform its thus accurately determined course; and then,

* So that the Atlantic tidal wave, as a wave, is equatorial in its origin, and passes round the coast of America, and so across the Atlantic to Europe.

† The co-tidal lines of the Atlantic have been drawn under the impression that the Atlantic tidal wave is, in each instance, a single wave, which, starting from the Antarctic regions, passes between Africa and South America, simultaneously causing a tide on each of these coasts; then crosses the equator, and, spreading over the North Atlantic, carries the flood-tide with it to Europe on the one side, and North America on the other. Moreover, they have been adapted to what Professor Airy has styled "the miserable equilibrium theory;" so that even he does not consider them very reliable: indeed, he says, "it is impossible to receive one of Mr. Whewell's speculations on the tides of the Atlantic." See *Tides and Waves.*

through comparing the actual space traversed in this period with that which has to be traversed by the other waves, in this way to arrive at an approximation of the probable period which these would occupy in passing through their more extended course. In doing this, the first thing to be noticed is that the polar waves, from having the same linear areas to traverse, will occupy the same periods respectively,* and therefore will meet at their common point of reflection, so that in reality the relations of only two waves have to be considered,—the equatorial and the polar; and here, for the first time, the probable origin of the semi-diurnal tide is foreshadowed.†

Of these two waves, the receding or equatorial, as being the shortest, the least modified, and at the same time the one whose origin can be most readily referred to the position and direct action of the moon, which it simply follows, offers the readiest data for the solution of the problem under examination.

Now, in the Atlantic Ocean, this wave has a determinate distance equal to a mean of some 45° to traverse; for the distance between the continents of Africa and South America —say between Cape St. Catherine and Cape St. Roque— is about 45°. But 45° are equal to three hours of mean time, as measured by the earth's axial rotation. Hence the moon's period, or the time which it takes in passing from the one continent to the other, is thus found to be three hours. The time occupied by the receding wave can also, and as readily, be learnt; for at the full and new

* The *ovoid* figure of the earth will give the southern polar wave a greater surface to traverse than the northern has to pass over; but on the whole area this difference will be relatively slight, and need not be regarded here.

† When it has been recognised and admitted that the semi-diurnal tides originate in these waves, it will be further seen that the diurnal tide is caused either by one of these receiving an undue prominence in range, or else from a reflected and a direct wave meeting each other, but not absolutely coinciding.

moon, when the moon crosses the meridian of St. Roque at about two o'clock of Greenwich mean time, the tide is not high at that point till six; so that the receding or equatorial wave does not culminate there till four hours after the moon has crossed the meridian, or seven hours after it left its meridian of genesis at the west coast of Africa.*

These relations are most important, and deserve very careful study; for through them the Atlantic equatorial regions can be made to the oceanic tides what the base line of a triangle is to its sides in the geometrical measurement of distance—a starting-point from which to determine their relations to each other.

The problem sought to be solved is this: admitting that one of the semi-diurnal tides is equatorial in its origin, is it possible that the other is derived from the polar waves? If it is, then the semi-diurnal tides being, speaking in round numbers, about twelve hours apart, the polar tides, starting from the same meridian at the same time, should reach a determinate position on the equator twelve hours after the equatorial wave. Hence, since the Atlantic equatorial wave reaches Cape St. Roque four hours after the moon, then the polar waves should arrive there yet twelve hours later. The issue raised, therefore, is this: Is it probable—nay, possible—that the polar waves would occupy this time?

To answer this question with approximate correctness, the meridian of Greenwich offers itself, by position, as the best point of departure in the relative calculations. Now, the meridian of Greenwich is some 35° east of the meridian of Cape St. Roque, or, in round numbers, some two hours in mean Greenwich time;† but the tidal wave takes six hours to

* The same ratio is observed on the east coast of Africa, where this tidal wave also follows the moon after an interval of about 4 hours—equal to some 60° of longitude.

† The meridian of Greenwich offers itself as a convenient point of depar-

traverse this distance, or three times as long as the moon. Then, again, the distance to be traversed by the polar wave is about three times that passed over by the equatorial. It ought, therefore, on a *primâ facie* view, to take thrice the time, or eighteen hours; and so reach Cape St. Roque some twelve hours after it,—which is the interval between the semi-diurnal tides.

A more careful analysis will only confirm the strict accord between the results determined by observation and required by theory; for the equatorial distance is some 35° (not 30°), and therefore represents 2^h 20^m in mean time; so that six hours is one hour less than thrice that: but then the moon does not commence its action till after it has crossed the meridian, and has reached a commensurate angular distance from it, and then with a minimum of relative attracting force; and, on the other hand, the curved path of the polar wave is necessarily longer than 90°, probably equal to three times 35° or 105°; so that the polar wave is initiated earlier than the equatorial, and by a maximum, the equatorial beginning later, and with a minimum, of the moon's attraction. Hence, while on the one hand the moon's position renders three times its own time necessary to the motion of the equatorial waters, it on the other causes three times the time to be sufficient for the greater relative linear area traversed by the polar wave, by imparting a higher velocity to that wave as it advances. But beyond this, another very important element—*revulsion*, caused by the action of the gravity of the earth to preserve its individual equilibrium

ture in this calculation; but disregarding this, and only considering the geometrical relations of the culminating wave to the meridian over which the moon is passing, the 60° interval represents that point in the angular relations of the relative attractions of the moon and the earth, on the surface of the latter, at which the moon exercises its greatest disturbing influence; so that the crest of the equatorial or receding wave probably always follows the moon at this interval, but yet requires an extensive coast on which to impinge, to make itself sensible in its true relations.

as a spheroid of revolution, and with this the stability of its polar axis—aids in generating and maintaining the actual synchronism of the observed tidal waves, and causes an interval in time, equal to half a sphere in linear distance, to intervene between them.*

If this is the case—and that it is the case can hardly be doubted, since no other satisfactory way of accounting for semi-diurnal tides seems to be attainable—then it becomes evident that disturbing bodies, acting over the earth's equator, simultaneously initiate two classes of tidal waves—*equatorial* and *polar*—which, starting from a given meridian at the same time, though 90° apart, reach a determinate point on the equator some twelve hours after each other, and are then propagated from the equator towards the poles as the observed tides, following each in the track of the other, with the same interval persisting between them.†

Some significant results flow from this induction; for, if the polar wave reaches the equator some eighteen hours after the moon has left the meridian from which it started, it follows that when it reaches the equator, though eighteen hours behind the moon in fact, it is six hours in advance of it in position, since the moon will now be on a meridian 90° to the east of it. Hence, during the last twelve hours of its course, as a direct wave, it has been unaided by the moon's

* If the earth is so poised in space that its northern hemisphere gravitates to the north celestial pole, then, under whatever circumstances the tidal waves may originate, the revulsive tendency of the earth, generated by its individual gravity, and indicated by the stability of its polar axis, would cause an interval equal to an hemisphere to intervene between them: as, if a single wave were passing round the globe, or two waves in nearer proximity than 180°, the polar axis would oscillate round its centre or vary its position; this is obviated by the law of revulsion, which, preserving the interval of 180° between them, as shown by the actual synchronism of the tides, causes the one to counterpoise the other, and so maintains the stability of the axis of revolution.

† The admirable manner in which this arrangement, by preserving the synchronous relations of the tides, aids in maintaining the stability of the earth's polar axis, is obvious.

attraction—as it will continue during the remainder of its existence as a wave, since the moon's direct influence, once having ceased, cannot be resumed on that individual wave, which is now subject solely to the laws of projection, reflection, and equilibrium. And hence, when it reaches the coast of England as a reflected wave some twenty-four hours later, although the "establishment of each port" is determined by the then position of the moon, and, owing to the laws governing the earth's equilibrium, can be so computed with exactness, nevertheless the actual wave has no other relations than those determined by the laws of equilibrium to the then position of the moon, seeing that its relations with that body are only and simply these: that it was set in motion by it some two revolutions earlier; so that, although the influence of the moon in determining the tides was primarily learned by the persistent relations of the establishment of the given port to its phases, it is now seen that a just inference was drawn from a mistaken interpretation, itself originating in a coincidence, in which either phenomenon only stood indirectly to the other in the relation of cause and effect.

Accepting this theory as a probably accurate explanation of the observed tides, many of the hitherto unexplained tidal phenomena will receive a ready solution: thus, the highest spring-tide and the lowest neap would necessarily follow the respective phases of the moon at an interval equivalent to the one recognised under observation, because the wave would occupy that time in moving from its point of genesis round the coast of North America and across the Atlantic. Then the slight range of the tides in the equatorial, as compared with that in the temperate, or circumpolar regions, is intelligible; since the former is a direct or diffused, the latter a reflected, and therefore more or less concentrated, wave. And further, the great range of the derived tides in channels and rivers is accounted for, seeing that it depends upon varying degrees of concentration, and the direction of the mouth and general course, with reference to the special wave from which that under examination is derived.

The theory of the tides in channels or canals has been already fully discussed by the present Astronomer Royal,* it need not therefore be considered here; but in examining it, care must be taken in determining from which wave the tide under review has been most probably derived. Thus that in the Amazon will originate alternately in the equatorial and polar direct waves; that in the St. Lawrence will be a lateral divergence from the great reflected wave moving in the course of the Gulf Stream; while that in the English and Irish channels, and the Severn, &c., will be the direct expression of this reflected wave.

In the same manner, the comparative freedom from tides found in the Pacific can be accounted for, since here there is no equivalent for the reflecting continent of South America, the mass of islands found in their course tending rather to break up the direct (*diffused*) waves; while the reacting waves, which are also diffused, are able, for the most part, to cross each other, and pass on in their original directions into the opposite hemisphere, until concentration makes them sensible. But even here the sources of the derived wave can

* It is not a little singular that the very basis of Professor Airy's wave theory, drawn from the observation of *short* waves, is, that in the hollow of the wave, after its crest has passed, the water is retreating, or flowing contrary to the direction of the wave—that is to say, that its stream moves towards the lowest level of the water. In *long* waves this is not the case, for then, as in the English Channel, the flow continues in the direction of the wave after its crest has passed; since the water runs up channel some three hours after its height is falling, or until the mean range has been reached. The same holds good of the ebb tide, when the flow continues against the rising water.

Professor Airy has also left the phenomenon of stationary waves unnoticed, such as those seen in "Races" and the like. These invert the relations of moving waves; the broken crest, as well as of course the steepest side, being directed towards, or meeting, the flowing water. This condition it is that makes the passage of races dangerous, since the stream carries the vessel into and under the opposed broken water.

generally be indicated—as, in the Gulf of California, from the alternate equatorial and South Polar reacting waves—from which also the tides in India and the Persian and Arabian Gulfs will be chiefly derived, those in China, Japan, &c., being, probably in part, aided by reflection;* but indeed, owing to the land being chiefly massed in the northern hemisphere, the diffused Pacific waves have but little facility for concentration, and will, in great part, be equatorial and antarctic in their origin. Even on the east coast of Africa they will probably be rather direct than reflected in their sensible effects,† while on its west coast they will be expressions of the direct antarctic wave.‡ So little, however, is known of the tides in the Pacific, that no generalization can be made upon them with any degree of reliability, except that its waters are so much more diffused than those of the Atlantic, that diffusion will most probably, unless in exceptional cases, be the principal characteristic of its tides.§

In the north of Europe again, and especially in the North Sea, it is possible that the direct wave coming from the Arctic regions may meet, and apparently combine, with the reflex wave arriving from the coast of North America.‖

* The combinations of the so-called diurnal with the semi-diurnal wave, as at Petropaulovski, which give the semblance of a single tide in the day, will arise from a direct or a reacting wave intervening between two tides, and so preventing one of the low waters by rising as the other is falling.

† The interval between the passage of the moon across the meridian and the arrival of the crest of the wave—some four hours—appears to indicate this.

‡ Diffused offshoots of it.

§ Want of coincidence in the diffused waves and imperfect reflection and concentration are probably the cause of the slight range and seeming irregularities of the tides in the Pacific; accidental, or rather intermitting causes, sometimes leading to the predominance of the one, and sometimes of the other, of the three principal direct waves.

‖ What has been termed the diurnal tide is, in reality, a want of exact coincidence in the simultaneously moving waves. Though it is possible that in some regions, or under certain conditions, a revulsive wave may be

If these inductions are true, and they certainly offer a reasonable and probable explanation of the several recognised tidal phenomena, then it must be admitted that, instead of the semi-diurnal tides observed in Europe being alternately a single wave following the moon, or a double one following the moon and sun, they are, in each instance, a reflected wave, and point to at least six classes of waves as being in operation in producing them.

1st. The *direct*—diffused waves, passing from the poles towards the equator.

2nd. The *receding*—a diffused wave, passing westwards in the equatorial regions.

3rd. The *reacting*—diffused waves, passing from the equator towards the poles.

4th. The *reflected*—concentrated waves, in the North Atlantic, passing from the east coast of America to the west coast of Europe.

5th. The *derived*—concentrated waves, passing into the several channels and rivers diverging from the ocean; and,

6th. The *synchronous* or *revulsive*—not distinct waves, but rather a controlling and guiding tendency, imparting a periodicity to any and all of the other waves, in order thus to maintain the equilibrium and general stability of the terrestrial globe in its variable relations.

But then, again, these six classes of waves are resolvable into two—the equatorial and the polar—which, as reflected from the equatorial regions, alternately form the semi-diurnal tides recognised under observation on the coasts of, or that diverge from, the North Atlantic Ocean; so that, if this theory is true, the North Atlantic semi-diurnal tides are derived alternately from diffused waves, originating simultaneously in the equatorial and polar regions of the earth, which, owing to the conditions and circumstances of their

generated, to preserve the equilibrium of the earth and the stability of its polar axis as a spheroid of revolution.

genesis, reach the north-east coast of South America some twelve hours after each other, and are then reflected, as concentrated waves, round the east coast of North America and across the Atlantic Ocean to Europe, constituting in this part of their course the observed tides.

London: Benjamin Pardon, Printer, Paternoster Row.

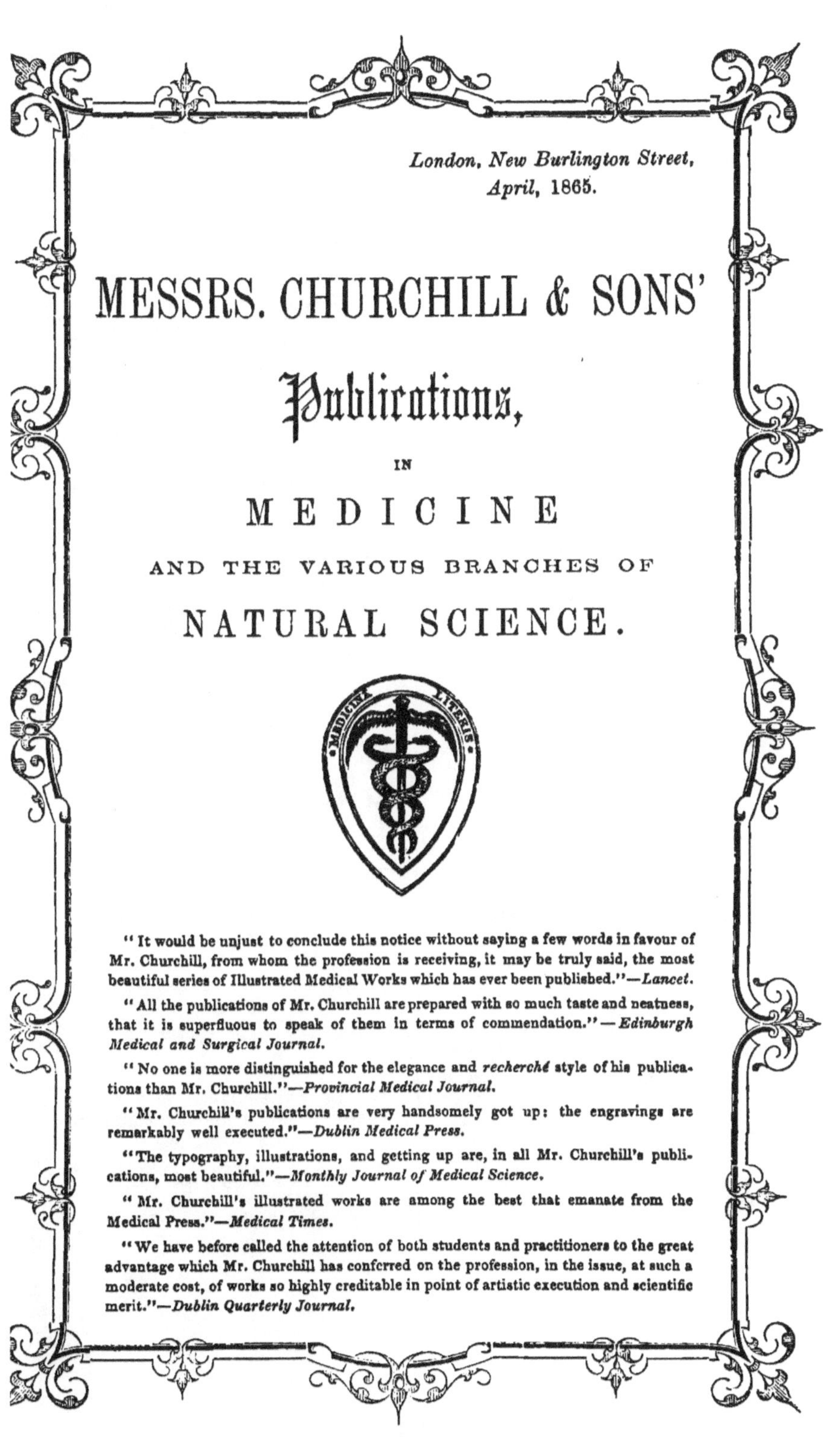

London, New Burlington Street,
April, 1865.

MESSRS. CHURCHILL & SONS'

Publications,

IN

MEDICINE

AND THE VARIOUS BRANCHES OF

NATURAL SCIENCE.

" It would be unjust to conclude this notice without saying a few words in favour of Mr. Churchill, from whom the profession is receiving, it may be truly said, the most beautiful series of Illustrated Medical Works which has ever been published."—*Lancet.*

" All the publications of Mr. Churchill are prepared with so much taste and neatness, that it is superfluous to speak of them in terms of commendation."—*Edinburgh Medical and Surgical Journal.*

" No one is more distinguished for the elegance and *recherché* style of his publications than Mr. Churchill."—*Provincial Medical Journal.*

" Mr. Churchill's publications are very handsomely got up: the engravings are remarkably well executed."—*Dublin Medical Press.*

" The typography, illustrations, and getting up are, in all Mr. Churchill's publications, most beautiful."—*Monthly Journal of Medical Science.*

" Mr. Churchill's illustrated works are among the best that emanate from the Medical Press."—*Medical Times.*

" We have before called the attention of both students and practitioners to the great advantage which Mr. Churchill has conferred on the profession, in the issue, at such a moderate cost, of works so highly creditable in point of artistic execution and scientific merit."—*Dublin Quarterly Journal.*

MESSRS. CHURCHILL & SONS are the Publishers of the following Periodicals, offering to Authors a wide extent of Literary Announcement, and a Medium of Advertisement, addressed to all Classes of the Profession.

THE BRITISH AND FOREIGN MEDICO-CHIRURGICAL REVIEW, AND QUARTERLY JOURNAL OF PRACTICAL MEDICINE AND SURGERY.

Price Six Shillings. Nos. I. to LXX.

THE QUARTERLY JOURNAL OF SCIENCE.

Price Five Shillings. Nos. I. to VI.

THE QUARTERLY JOURNAL OF MICROSCOPICAL SCIENCE,

INCLUDING THE TRANSACTIONS OF THE MICROSCOPICAL SOCIETY OF LONDON.

Edited by DR. LANKESTER, F.R.S., and GEORGE BUSK, F.R.S. Price 4s. Nos. I. to XVIII. *New Series.*

THE JOURNAL OF MENTAL SCIENCE.

By authority of the Association of Medical Officers of Asylums and Hospitals for the Insane. Edited by C. L. ROBERTSON, M.D., and HENRY MAUDSLEY, M.D.

Published Quarterly, price Half-a-Crown. *New Series.* Nos. I. to XVII.

THE JOURNAL OF BRITISH OPHTHALMOLOGY AND QUARTERLY REPORT OF OPHTHALMIC MEDICINE AND SURGERY.

Edited by JABEZ HOGG, Surgeon. Price 2s. 6d. No. I.

ARCHIVES OF MEDICINE:

A Record of Practical Observations and Anatomical and Chemical Researches, connected with the Investigation and Treatment of Disease. Edited by Dr. LIONEL S. BEALE, F.R.S. Published Quarterly; Nos. I. to VIII., 3s. 6d.; IX. to XII., 2s. 6d., XIII., XIV., 3s.

ARCHIVES OF DENTISTRY:

Edited by EDWIN TRUMAN. Published Quarterly, price 4s. Nos. I. & II.

THE ROYAL LONDON OPHTHALMIC HOSPITAL REPORTS, AND JOURNAL OF OPHTHALMIC MEDICINE AND SURGERY.

Vol. IV., Part 3, 2s. 6d.

THE MEDICAL TIMES & GAZETTE.

Published Weekly, price Sixpence, or Stamped, Sevenpence.

Annual Subscription, £1. 6s., or Stamped, £1. 10s. 4d., and regularly forwarded to all parts of the Kingdom.

THE HALF-YEARLY ABSTRACT OF THE MEDICAL SCIENCES.

Being a Digest of the Contents of the principal British and Continental Medical Works; together with a Critical Report of the Progress of Medicine and the Collateral Sciences. Post 8vo. cloth, 6s. 6d. Vols. I. to XL.

THE PHARMACEUTICAL JOURNAL,

CONTAINING THE TRANSACTIONS OF THE PHARMACEUTICAL SOCIETY.

Published Monthly, price One Shilling.

*** Vols. I. to XXII., bound in cloth, price 12s. 6d. each.

THE BRITISH JOURNAL OF DENTAL SCIENCE.

Published Monthly, price One Shilling. Nos. I. to CV.

THE MEDICAL DIRECTORY FOR THE UNITED KINGDOM.

Published Annually. 8vo. cloth, 10s. 6d.

ANNALS OF MILITARY AND NAVAL SURGERY AND TROPICAL MEDICINE AND HYGIENE,

Embracing the experience of the Medical Officers of Her Majesty's Armies and Fleets in all parts of the World.

Published Annually. Vol. I., price 7s.

A CLASSIFIED INDEX
TO
MESSRS. CHURCHILL & SONS' CATALOGUE.

ANATOMY.

PAGE

Anatomical Remembrancer .. 3
Flower on Nerves 11
Hassall's Micros. Anatomy .. 14
Heale's Anatomy of the Lungs 14
Heath's Practical Anatomy .. 15
Holden's Human Osteology .. 15
Do. on Dissections 15
Huxley's Comparative Anatomy 16
Jones' and Sieveking's Pathological Anatomy 17
Maclise's Surgical Anatomy .. 19
St. Bartholomew's Hospital Catalogue 24
Sibson's Medical Anatomy .. 25
Waters' Anatomy of Lung .. 29
Wheeler's Anatomy for Artists 30
Wilson's Anatomy 31

CHEMISTRY.

Abel & Bloxam's Handbook .. 3
Bowman's Practical Chemistry 7
Do. Medical do. .. 7
Fownes' Manual of Chemistry.. 12
Do. Actonian Prize 12
Do. Qualitative Analysis .. 12
Fresenius' Chemical Analysis.. 12
Galloway's First Step 12
Do. Second Step 12
Do. Analysis 12
Do. Tables 12
Griffiths' Four Seasons 13
Horsley's Chem. Philosophy .. 16
Mulder on the Chemistry of Wine 20
Plattner & Muspratt on Blowpipe 22
Speer's Pathol. Chemistry .. 26
Sutton's Volumetric Analysis . 27

CLIMATE.

Barker on Worthing 4
Bennet on Mentone 6
Dalrymple on Egypt.. 10
Francis on Change of Climate.. 12
Hall on Torquay.. 14
Haviland on Climate.. 14
Lee on Climate 18
Do. Watering Places of England 18
McClelland on Bengal 19
McNicoll on Southport 19
Martin on Tropical Climates .. 20
Moore's Diseases of India .. 20
Scoresby-Jackson's Climatology 24
Shapter on South Devon.. .. 25
Siordet on Mentone 25
Taylor on Pau and Pyrenees .. 27

DEFORMITIES, &c.

Adams on Spinal Curvature .. 3
Barwell on Clubfoot 4
Bigg on Deformities 6
Do. on Artificial Limbs 6
Bishop on Deformities 6
Do. Articulate Sounds .. 6
Brodhurst on Spine 7
Do. on Clubfoot 7
Godfrey on Spine 13
Hugman on Hip Joint 16
Tamplin on Spine 27

DISEASES OF WOMEN AND CHILDREN.

PAGE

Ballard on Infants and Mothers 4
Bennet on Uterus 6
Do. on Uterine Pathology.. 6
Bird on Children 7
Bryant's Surgical Diseases of Children 7
Eyre's Practical Remarks .. 11
Harrison on Children 14
Hood on Scarlet Fever, &c. .. 16
Kiwisch (ed. by Clay) on Ovaries 9
Lee's Ovarian & Uterine Diseases 18
Do. on Diseases of Uterus .. 18
Do. on Speculum 18
Seymour on Ovaria 25
Smith on Leucorrhœa 26
Tilt on Uterine Inflammation.. 28
Do. Uterine Therapeutics .. 28
Do. on Change of Life 28
Underwood on Children 29
Wells on the Ovaries.. 30
West on Women 30
Do. (Uvedale) on Puerperal Diseases 30

GENERATIVE ORGANS, Diseases of, and SYPHILIS.

Acton on Reproductive Organs 3
Coote on Syphilis 10
Gant on Bladder.. 13
Hutchinson on Inherited Syphilis 16
Judd on Syphilis 17
Lee on Syphilis 18
Parker on Syphilis 21
Wilson on Syphilis 31

HYGIENE.

Armstrong on Naval Hygiene 4
Beale's Laws of Health 5
Do. Health and Disease .. 5
Bennet on Nutrition 6
Carter on Training 8
Chavasse's Advice to a Mother.. 9
Do. Advice to a Wife .. 9
Dobell's Germs and Vestiges of Disease 11
Do. Diet and Regimen .. 11
Granville on Vichy 13
Hartwig on Sea Bathing 14
Do. Physical Education 14
Hufeland's Art of prolonging Life 16
Lee's Baths of Germany 18
Moore's Health in Tropics .. 20
Parkes on Hygiene 21
Parkin on Disease 21
Pickford on Hygiene 21
Robertson on Diet 24
Routh on Infant Feeding.. .. 23
Rumsey's State Medicine 24
Tunstall's Bath Waters 28
Wells' Seamen's Medicine Chest 30
Wife's Domain 30
Wilson on Healthy Skin 31
Do. on Mineral Waters .. 31
Do. on Turkish Bath 31

MATERIA MEDICA and PHARMACY.

Bateman's Magnacopia 5
Beasley's Formulary 5
Do. Receipt Book 5
Do. Book of Prescriptions 5
Frazer's Materia Medica 12
Nevins' Analysis of Pharmacop. 20
Pereira's Selecta è Præscriptis 21

MATERIA MEDICA and PHARMACY—*continued.*

PAGE

Pharmacopœia Londinensis .. 22
Prescriber's Pharmacopœia .. 22
Royle's Materia Medica 24
Squire's Hospital Pharmacopœias 26
Do. Companion to the Pharmacopœia 26
Steggall's First Lines for Chemists and Druggists 26
Stowe's Toxicological Chart .. 27
Taylor on Poisons 27
Waring's Therapeutics 29
Wittstein's Pharmacy 31

MEDICINE.

Adams on Rheumatic Gout .. 3
Addison on Cell Therapeutics.. 3
Do. on Healthy and Diseased Structure 3
Aldis's Hospital Practice 3
Anderson on Fever 4
Austin on Paralysis 4
Barclay on Medical Diagnosis.. 4
Barlow's Practice of Medicine 4
Basham on Dropsy 5
Brinton on Stomach 7
Do. on Ulcer of do. 7
Budd on the Liver 8
Do. on Stomach 8
Camplin on Diabetes.. 8
Chambers on Digestion 8
Do. Lectures 8
Davey's Ganglionic Nervous System 11
Eyre on Stomach 11
French on Cholera 12
Fuller on Rheumatism 12
Gairdner on Gout 12
Gibb on Throat 13
Granville on Sudden Death .. 13
Gully's Simple Treatment .. 13
Habershon on the Abdomen .. 13
Do. on Mercury 13
Hall (Marshall) on Apnœa .. 14
Do. Observations .. 14
Headland—Action of Medicines 14
Hooper's Physician's Vade-Mecum 13
Inman's New Theory 16
Do. Myalgia.. 16
James on Laryngoscope 17
Maclachlan on Advanced Life.. 19
Marcet on Chronic Alcoholism . 19
Meryon on Paralysis 20
Pavy on Diabetes 21
Peacock on Influenza 21
Peet's Principles and Practice of Medicine 21
Richardson's Asclepiad 23
Roberts on Palsy 23
Robertson on Gout 24
Savory's Compendium 24
Semple on Cough 24
Seymour on Dropsy 25
Shaw's Remembrancer 25
Smee on Debility 25
Thomas' Practice of Physic .. 27
Thudichum on Gall Stones .. 28
Todd's Clinical Lectures 28
Tweedie on Continued Fevers 29
Walker on Diphtheria 29
Wells on Gout 30
What to Observe at the Bedside 19
Williams' Principles 30
Wright on Headaches 31

CLASSIFIED INDEX.

MICROSCOPE.

PAGE
Beale on Microscope in Medicine .. 5
Carpenter on Microscope 8
Schacht on do. 24

MISCELLANEOUS.

Acton on Prostitution 3
Barclay's Medical Errors 4
Bascome on Epidemics 4
Bryce on Sebastopol 8
Cooley's Cyclopædia 9
Gordon on China 13
Graves' Physiology and Medicine 13
Guy's Hospital Reports 13
Harrison on Lead in Water .. 14
Hingeston's Topics of the Day .. 15
Lane's Hydropathy 18
Lee on Homœop. and Hydrop. 18
London Hospital Reports 19
Marcet on Food 19
Massy on Recruits 20
Mayne's Medical Vocabulary .. 20
Part's Case Book 21
Redwood's Supplement to Pharmacopœia 23
Ryan on Infanticide 24
Snow on Chloroform 26
Steggall's Medical Manual .. 26
Do. Gregory's Conspectus 26
Do. Celsus 26
Whitehead on Transmission .. 30

NERVOUS DISORDERS AND INDIGESTION.

Birch on Constipation 6
Carter on Hysteria 8
Downing on Neuralgia 11
Hunt on Heartburn 16
Jones (Handfield) on Functional Nervous Disorders 17
Leared on Imperfect Digestion 18
Lobb on Nervous Affections .. 19
Radcliffe on Epilepsy 22
Reynolds on the Brain 23
Do. on Epilepsy 23
Rowe on Nervous Diseases .. 24
Sieveking on Epilepsy 25
Turnbull on Stomach 28

OBSTETRICS.

Barnes on Placenta Prævia .. 4
Hodges on Puerperal Convulsions 15
Lee's Clinical Midwifery 18
Do. Consultations 18
Leishman's Mechanism of Parturition 18
Mackenzie on Phlegmasia Dolens 19
Pretty's Aids during Labour .. 22
Priestley on Gravid Uterus .. 22
Ramsbotham's Obstetrics 23
Do. Midwifery 23
Sinclair & Johnston's Midwifery 25
Smellie's Obstetric Plates 25
Smith's Manual of Obstetrics .. 26
Swayne's Aphorisms 27
Waller's Midwifery 29

OPHTHALMOLOGY.

Cooper on Injuries of Eye .. 9
Do. on Near Sight 9
Dalrymple on Eye 10
Dixon on the Eye 11
Hogg on Ophthalmoscope .. 15
Holthouse on Strabismus 15
Do. on Impaired Vision 15

OPHTHALMOLOGY—*contd.*

PAGE
Hulke on the Ophthalmoscope 16
Jacob on Eye-ball 16
Jago on Entoptics 17
Jones' Ophthalmic Medicine .. 17
Do. Defects of Sight 17
Do. Eye and Ear 17
Nunneley on the Organs of Vision 21
Walton on the Eye 29
Wells on Spectacles 30

PHYSIOLOGY.

Carpenter's Human 8
Do. Comparative 8
Do. Manual 8
Heale on Vital Causes 14
O'Reilly on the Nervous System 21
Richardson on Coagulation .. 23
Shea's Animal Physiology 25
Virchow's (ed. by Chance) Cellular Pathology 8

PSYCHOLOGY.

Arlidge on the State of Lunacy 4
Bucknill and Tuke's Psychological Medicine 8
Conolly on Asylums 9
Davey on Nature of Insanity .. 11
Dunn's Physiological Psychology 11
Hood on Criminal Lunatics .. 16
Millingen on Treatment of Insane 20
Noble on Mind 21
Williams (J. H.) Unsoundness of Mind 30

PULMONARY and CHEST DISEASES, &c.

Alison on Pulmonary Consumption 3
Billing on Lungs and Heart .. 6
Bright on the Chest 7
Cotton on Consumption 10
Do. on Stethoscope 10
Davies on Lungs and Heart .. 10
Dobell on the Chest 11
Fenwick on Consumption 11
Fuller on Chest 12
Do. on Heart 12
Jones (Jas.) on Consumption .. 17
Laennec on Auscultation 18
Markham on Heart 20
Richardson on Consumption .. 23
Salter on Asthma 24
Skoda on Auscultation 20
Thompson on Consumption .. 27
Timms on Consumption 28
Turnbull on Consumption .. 28
Waters on Emphysema 29
Weber on Auscultation 29

RENAL and URINARY DISEASES.

Acton on Urinary Organs .. 3
Beale on Urine 5
Bird's Urinary Deposits 6
Coulson on Bladder 10
Hassall on Urine 14
Parkes on Urine 21
Thudichum on Urine 28
Todd on Urinary Organs 28

SCIENCE.

Baxter on Organic Polarity .. 5
Bentley's Manual of Botany .. 6
Bird's Natural Philosophy .. 6

SCIENCE—*continued.*

PAGE
Craig on Electric Tension .. 10
Hardwich's Photography 14
Hinds' Harmonies 15
Howard on the Clouds 16
Jones on Vision 17
Do. on Body, Sense, and Mind 17
Mayne's Lexicon 20
Pratt's Genealogy of Creation .. 22
Do. Eccentric & Centric Force 22
Do. on Orbital Motion 22
Price's Photographic Manipulation 22
Rainey on Shells 23
Reymond's Animal Electricity 23
Taylor's Medical Jurisprudence 27
Unger's Botanical Letters .. 29
Vestiges of Creation 29

SURGERY.

Adams on Reparation of Tendons 3
Do. Subcutaneous Surgery 3
Anderson on the Skin 3
Ashton on Rectum 4
Barwell on Diseases of Joints .. 4
Brodhurst on Anchylosis 7
Bryant on Diseases of Joints .. 7
Callender on Rupture 8
Chapman on Ulcers 9
Do. Varicose Veins 9
Clark's Outlines of Surgery .. 9
Colles on Cancer 9
Cooper (Sir A.) on Testis 10
Do. (S.) Surg. Dictionary 10
Coulson on Lithotomy 10
Curling on Rectum 10
Do. on Testis 10
Druitt's Surgeon's Vade-Mecum 11
Fergusson's Surgery 11
Gant's Principles of Surgery .. 13
Heath's Minor Surgery and Bandaging 15
Higginbottom on Nitrate of Silver 15
Hodgson on Prostate 15
Holt on Stricture 15
James on Hernia 17
Jordan's Clinical Surgery .. 17
Lawrence's Surgery 18
Do. Ruptures 18
Liston's Surgery 18
Logan on Skin Diseases 19
Macleod's Surgical Diagnosis .. 19
Do. Surgery of the Crimea 19
Maclise on Fractures 19
Maunder's Operative Surgery .. 20
Nunneley on Erysipelas 21
Pirrie's Surgery 22
Savage's Female Pelvic Organs 24
Smith (Hy.) on Stricture 25
Do. on Hæmorrhoids 25
Do. (Dr. J.) Dental Anatomy and Surgery 26
Squire on Skin Diseases 26
Steggall's Surgical Manual .. 26
Teale on Amputation 27
Thompson on Stricture 27
Do. on Prostate 27
Do. Lithotomy and Lithotrity 27
Tomes' Dental Surgery 28
Toynbee on Ear 28
Wade on Stricture 29
Watson on the Larynx 29
Webb's Surgeon's Ready Rules 29
Williamson on Military Surgery 30
Do. on Gunshot Injuries 30
Wilson on Skin Diseases 31
Do. Portraits of Skin Diseases 31
Yearsley on Deafness 31
Do. on Throat 31

MR. F. A. ABEL, F.R.S., & MR. C. L. BLOXAM.

HANDBOOK OF CHEMISTRY: THEORETICAL, PRACTICAL, AND TECHNICAL. Second Edition. 8vo. cloth, 15*s.*

MR. ACTON, M.R.C.S.

I.

A PRACTICAL TREATISE ON DISEASES OF THE URINARY AND GENERATIVE ORGANS IN BOTH SEXES. Third Edition. 8vo. cloth, £1. 1*s.* With Plates, £1. 11*s.* 6*d.* The Plates alone, limp cloth, 10*s.* 6*d.*

II.

THE FUNCTIONS AND DISORDERS OF THE REPRODUCTIVE ORGANS IN CHILDHOOD, YOUTH, ADULT AGE, AND ADVANCED LIFE, considered in their Physiological, Social, and Moral Relations. Fourth Edition. 8vo. cloth, 10*s.* 6*d.*

III.

PROSTITUTION: Considered in its Moral, Social, and Sanitary Bearings, with a View to its Amelioration and Regulation. 8vo. cloth, 10*s.* 6*d.*

DR. ADAMS, A.M.

A TREATISE ON RHEUMATIC GOUT; OR, CHRONIC RHEUMATIC ARTHRITIS. 8vo. cloth, with a Quarto Atlas of Plates, 21*s.*

MR. WILLIAM ADAMS, F.R.C.S.

I.

ON THE PATHOLOGY AND TREATMENT OF LATERAL AND OTHER FORMS OF CURVATURE OF THE SPINE. With Plates. 8vo. cloth, 10*s.* 6*d.*

II.

ON THE REPARATIVE PROCESS IN HUMAN TENDONS AFTER SUBCUTANEOUS DIVISION FOR THE CURE OF DEFORMITIES. With Plates. 8vo. cloth, 6*s.*

III.

SKETCH OF THE PRINCIPLES AND PRACTICE OF SUBCUTANEOUS SURGERY. 8vo. cloth, 2*s.* 6*d.*

DR. WILLIAM ADDISON, F.R.S.

I.

CELL THERAPEUTICS. 8vo. cloth, 4*s.*

II.

ON HEALTHY AND DISEASED STRUCTURE, AND THE TRUE PRINCIPLES OF TREATMENT FOR THE CURE OF DISEASE, ESPECIALLY CONSUMPTION AND SCROFULA, founded on MICROSCOPICAL ANALYSIS. 8vo. cloth, 12*s.*

DR. ALDIS.

AN INTRODUCTION TO HOSPITAL PRACTICE IN VARIOUS COMPLAINTS; with Remarks on their Pathology and Treatment. 8vo. cloth, 5*s.* 6*d.*

DR. SOMERVILLE SCOTT ALISON, M.D.EDIN., F.R.C.P.

THE PHYSICAL EXAMINATION OF THE CHEST IN PULMONARY CONSUMPTION, AND ITS INTERCURRENT DISEASES. With Engravings. 8vo. cloth, 12*s.*

THE ANATOMICAL REMEMBRANCER; OR, COMPLETE POCKET ANATOMIST. Fifth Edition, carefully Revised. 32mo. cloth, 3*s.* 6*d.*

DR. MCCALL ANDERSON, M.D.

I.

PARASITIC AFFECTIONS OF THE SKIN. With Engravings. 8vo. cloth, 5*s.*

II.

PRACTICAL TREATISE ON ECZEMA. With Engravings. 8vo. cloth, 5*s.*

DR. ANDREW ANDERSON, M.D.

TEN LECTURES INTRODUCTORY TO THE STUDY OF FEVER. Post 8vo. cloth, 5*s.*

DR. ARLIDGE.

ON THE STATE OF LUNACY AND THE LEGAL PROVISION FOR THE INSANE; with Observations on the Construction and Organisation of Asylums. 8vo. cloth, 7*s.*

DR. ALEXANDER ARMSTRONG, R.N.

OBSERVATIONS ON NAVAL HYGIENE AND SCURVY. More particularly as the latter appeared during a Polar Voyage. 8vo. cloth, 5*s.*

MR. T. J. ASHTON.

I.

ON THE DISEASES, INJURIES, AND MALFORMATIONS OF THE RECTUM AND ANUS. Fourth Edition. 8vo. cloth, 8*s.*

II.

PROLAPSUS, FISTULA IN ANO, AND HÆMORRHOIDAL AFFECTIONS; their Pathology and Treatment. Second Edition. Post 8vo. cloth, 2*s.* 6*d.*

MR. THOS. J. AUSTIN, M.R.C.S.ENG.

A PRACTICAL ACCOUNT OF GENERAL PARALYSIS: Its Mental and Physical Symptoms, Statistics, Causes, Seat, and Treatment. 8vo. cloth, 6*s.*

DR. THOMAS BALLARD, M.D.

A NEW AND RATIONAL EXPLANATION OF THE DISEASES PECULIAR TO INFANTS AND MOTHERS; with obvious Suggestions for their Prevention and Cure. Post 8vo. cloth, 4*s.* 6*d.*

DR. BARCLAY.

I.

A MANUAL OF MEDICAL DIAGNOSIS. Second Edition. Foolscap 8vo. cloth, 8*s.* 6*d.*

II.

MEDICAL ERRORS.—Fallacies connected with the Application of the Inductive Method of Reasoning to the Science of Medicine. Post 8vo. cloth, 5*s.*

DR. W. G. BARKER.

ON THE CLIMATE OF WORTHING: its Remedial Influence in Disease, especially of the Lungs. Crown 8vo. cloth, 3*s.*

DR. BARLOW.

A MANUAL OF THE PRACTICE OF MEDICINE. Second Edition. Fcap. 8vo. cloth, 12*s.* 6*d.*

DR. BARNES.

THE PHYSIOLOGY AND TREATMENT OF PLACENTA PRÆVIA; being the Lettsomian Lectures on Midwifery for 1857. Post 8vo. cloth, 6*s.*

MR. BARWELL, F.R.C.S.

I.

A TREATISE ON DISEASES OF THE JOINTS. With Engravings. 8vo. cloth, 12*s.*

II.

ON THE CURE OF CLUBFOOT WITHOUT CUTTING TENDONS, and on certain new Methods of Treating other Deformities. With Engravings. Fcap. 8vo. cloth, 3*s.* 6*d.*

DR. BASCOME.

A HISTORY OF EPIDEMIC PESTILENCES, FROM THE EARLIEST AGES. 8vo. cloth, 8*s.*

DR. BASHAM.

I.

ON DROPSY, CONNECTED WITH DISEASE OF THE KIDNEYS (MORBUS BRIGHTII), and on some other Diseases of those Organs, associated with Albuminous and Purulent Urine. Illustrated by numerous Drawings from the Microscope. Second Edition. 8vo. cloth, 9*s.*

II.

THE SIGNIFICANCE OF DROPSY AS A SYMPTOM IN RENAL, CARDIAC, AND PULMONARY DISEASES. The Croonian Lectures for 1864. With Plates. 8vo. cloth, 5*s.*

MR. H. F. BAXTER, M.R.C.S.L.

ON ORGANIC POLARITY; showing a Connexion to exist between Organic Forces and Ordinary Polar Forces. Crown 8vo. cloth, 5*s.*

MR. BATEMAN.

MAGNACOPIA: A Practical Library of Profitable Knowledge, communicating the general Minutiæ of Chemical and Pharmaceutic Routine, together with the generality of Secret Forms of Preparations. Third Edition. 18mo. 6*s.*

MR. LIONEL J. BEALE, M.R.C.S.

I.

THE LAWS OF HEALTH IN THEIR RELATIONS TO MIND AND BODY. A Series of Letters from an Old Practitioner to a Patient. Post 8vo. cloth, 7*s.* 6*d.*

II.

HEALTH AND DISEASE, IN CONNECTION WITH THE GENERAL PRINCIPLES OF HYGIENE. Fcap. 8vo., 2*s.* 6*d.*

DR. BEALE, F.R.S.

I.

URINE, URINARY DEPOSITS, AND CALCULI: and on the Treatment of Urinary Diseases. Numerous Engravings. Second Edition, much Enlarged. Post 8vo. cloth, 8*s.* 6*d.*

II.

THE MICROSCOPE, IN ITS APPLICATION TO PRACTICAL MEDICINE. With a Coloured Plate, and 270 Woodcuts. Second Edition. 8vo. cloth, 14*s.*

III.

ILLUSTRATIONS OF THE SALTS OF URINE, URINARY DEPOSITS, and CALCULI. 37 Plates, containing upwards of 170 Figures copied from Nature, with descriptive Letterpress. 8vo. cloth, 9*s.* 6*d.*

MR. BEASLEY.

I.

THE BOOK OF PRESCRIPTIONS; containing 3000 Prescriptions. Collected from the Practice of the most eminent Physicians and Surgeons, English and Foreign. Third Edition. 18mo. cloth, 6*s.*

II.

THE DRUGGIST'S GENERAL RECEIPT-BOOK: comprising a copious Veterinary Formulary and Table of Veterinary Materia Medica; Patent and Proprietary Medicines, Druggists' Nostrums, &c.; Perfumery, Skin Cosmetics, Hair Cosmetics, and Teeth Cosmetics; Beverages, Dietetic Articles, and Condiments; Trade Chemicals, Miscellaneous Preparations and Compounds used in the Arts, &c.; with useful Memoranda and Tables. Fifth Edition. 18mo. cloth, 6*s.*

III.

THE POCKET FORMULARY AND SYNOPSIS OF THE BRITISH AND FOREIGN PHARMACOPŒIAS; comprising standard and approved Formulæ for the Preparations and Compounds employed in Medical Practice. Seventh Edition, corrected and enlarged. 18mo. cloth, 6*s.*

DR. HENRY BENNET.

I.

A PRACTICAL TREATISE ON INFLAMMATION AND OTHER DISEASES OF THE UTERUS. Fourth Edition, revised, with Additions. 8vo. cloth, 16*s.*

II.

A REVIEW OF THE PRESENT STATE OF UTERINE PATHOLOGY. 8vo. cloth, 4*s.*

III.

NUTRITION IN HEALTH AND DISEASE. Post 8vo. cloth, 5*s.*

IV.

MENTONE, THE RIVIERA, CORSICA, AND BIARRITZ, AS WINTER CLIMATES. Second Edition. Post 8vo. cloth, 5*s.*

PROFESSOR BENTLEY, F.L.S.

A MANUAL OF BOTANY. With nearly 1,200 Engravings on Wood. Fcap. 8vo. cloth, 12*s.* 6*d.*

MR. HENRY HEATHER BIGG.

I.

THE MECHANICAL APPLIANCES NECESSARY FOR THE TREATMENT OF DEFORMITIES.

Part I.—The Lower Limbs. Post 8vo. cloth, 4*s.*
Part II.—The Spine and Upper Extremities. Post 8vo. cloth, 4*s.* 6*d.*

II.

ARTIFICIAL LIMBS; THEIR CONSTRUCTION AND APPLICATION. With Engravings on Wood. 8vo. cloth, 3*s.*

DR. BILLING, F.R.S.

ON DISEASES OF THE LUNGS AND HEART. 8vo. cloth, 6*s.*

DR. S. B. BIRCH, M.D.

CONSTIPATED BOWELS: the Various Causes and the Rational Means of Cure. Second Edition. Post 8vo. cloth, 3*s.* 6*d.*

DR. GOLDING BIRD, F.R.S.

I.

URINARY DEPOSITS; THEIR DIAGNOSIS, PATHOLOGY, AND THERAPEUTICAL INDICATIONS. With Engravings. Fifth Edition. Edited by E. Lloyd Birkett, M.D. Post 8vo. cloth, 10*s.* 6*d.*

II.

ELEMENTS OF NATURAL PHILOSOPHY; being an Experimental Introduction to the Study of the Physical Sciences. With numerous Engravings. Fifth Edition. Edited by Charles Brooke, M.B. Cantab., F.R.S. Fcap. 8vo. cloth, 12*s.* 6*d.*

MR. BISHOP, F.R.S.

I.

ON DEFORMITIES OF THE HUMAN BODY, their Pathology and Treatment. With Engravings on Wood. 8vo. cloth, 10*s.*

II.

ON ARTICULATE SOUNDS, AND ON THE CAUSES AND CURE OF IMPEDIMENTS OF SPEECH. 8vo. cloth, 4*s.*

MR. P. HINCKES BIRD, F.R.C.S.

PRACTICAL TREATISE ON THE DISEASES OF CHILDREN AND INFANTS AT THE BREAST. Translated from the French of M. BOUCHUT, with Notes and Additions. 8vo. cloth. 20*s.*

MR. JOHN E. BOWMAN, & MR. C. L. BLOXAM.

I.

PRACTICAL CHEMISTRY, including Analysis. With numerous Illustrations on Wood. Fourth Edition. Foolscap 8vo. cloth, 6*s.* 6*d.*

II.

MEDICAL CHEMISTRY; with Illustrations on Wood. Fourth Edition, carefully revised. Fcap. 8vo. cloth, 6*s.* 6*d.*

DR. JAMES BRIGHT.

ON DISEASES OF THE HEART, LUNGS, & AIR PASSAGES; with a Review of the several Climates recommended in these Affections. Third Edition. Post 8vo. cloth, 9*s.*

DR. BRINTON, F.R.S.

I.

THE DISEASES OF THE STOMACH, with an Introduction on its Anatomy and Physiology; being Lectures delivered at St. Thomas's Hospital. Second Edition. 8vo. cloth, 10*s.* 6*d.*

II.

THE SYMPTOMS, PATHOLOGY, AND TREATMENT OF ULCER OF THE STOMACH. Post 8vo. cloth, 5*s.*

MR. BERNARD E. BRODHURST, F.R.C.S.

I.

CURVATURES OF THE SPINE: their Causes, Symptoms, Pathology, and Treatment. Second Edition. Roy. 8vo. cloth, with Engravings, 7*s.* 6*d.*

II.

ON THE NATURE AND TREATMENT OF CLUBFOOT AND ANALOGOUS DISTORTIONS involving the TIBIO-TARSAL ARTICULATION. With Engravings on Wood. 8vo. cloth, 4*s.* 6*d.*

III.

PRACTICAL OBSERVATIONS ON THE DISEASES OF THE JOINTS INVOLVING ANCHYLOSIS, and on the TREATMENT for the RESTORATION of MOTION. Third Edition, much enlarged, 8vo. cloth, 4*s.* 6*d.*

MR. THOMAS BRYANT, F.R.C.S.

I.

ON THE DISEASES AND INJURIES OF THE JOINTS. CLINICAL AND PATHOLOGICAL OBSERVATIONS. Post 8vo. cloth, 7*s.* 6*d.*

II.

THE SURGICAL DISEASES OF CHILDREN. The Lettsomian Lectures, delivered March, 1863. Post 8vo. cloth, 5*s.*

DR. BRYCE.

ENGLAND AND FRANCE BEFORE SEBASTOPOL, looked at from a Medical Point of View. 8vo. cloth, 6*s.*

DR. BUDD, F.R.S.

I.

ON DISEASES OF THE LIVER.

Illustrated with Coloured Plates and Engravings on Wood. Third Edition. 8vo. cloth, 16*s.*

II.

ON THE ORGANIC DISEASES AND FUNCTIONAL DISORDERS OF THE STOMACH. 8vo. cloth, 9*s.*

DR. JOHN CHARLES BUCKNILL, & DR. DANIEL H. TUKE.

A MANUAL OF PSYCHOLOGICAL MEDICINE: containing the History, Nosology, Description, Statistics, Diagnosis, Pathology, and Treatment of Insanity. Second Edition. 8vo. cloth, 15*s.*

MR. CALLENDER, F.R.C.S.

FEMORAL RUPTURE: Anatomy of the Parts concerned. With Plates. 8vo. cloth, 4*s.*

DR. JOHN M. CAMPLIN, F.L.S.

ON DIABETES, AND ITS SUCCESSFUL TREATMENT. Third Edition, by Dr. Glover. Fcap. 8vo. cloth, 3*s.* 6*d.*

MR. ROBERT B. CARTER, M.R.C.S.

I.

ON THE INFLUENCE OF EDUCATION AND TRAINING IN PREVENTING DISEASES OF THE NERVOUS SYSTEM. Fcap. 8vo., 6*s.*

II.

THE PATHOLOGY AND TREATMENT OF HYSTERIA. Post 8vo. cloth, 4*s.* 6*d.*

DR. CARPENTER, F.R.S.

I.

PRINCIPLES OF HUMAN PHYSIOLOGY. With numerous Illustrations on Steel and Wood. Sixth Edition. Edited by Mr. Henry Power. 8vo. cloth, 26*s.*

II.

PRINCIPLES OF COMPARATIVE PHYSIOLOGY. Illustrated with 300 Engravings on Wood. Fourth Edition. 8vo. cloth, 24*s.*

III.

A MANUAL OF PHYSIOLOGY. With 252 Illustrations on Steel and Wood. Fourth Edition. Fcap. 8vo. cloth, 12*s.* 6*d.*

IV.

THE MICROSCOPE AND ITS REVELATIONS. With numerous Engravings on Steel and Wood. Third Edition. Fcap. 8vo. cloth, 12*s.* 6*d.*

DR. CHAMBERS.

I.

LECTURES CHIEFLY CLINICAL. Third Edition, much enlarged. 8vo. cloth, 14*s.*

II.

DIGESTION AND ITS DERANGEMENTS. Post 8vo. cloth, 10*s.* 6*d.*

DR. CHANCE, M.B.

VIRCHOW'S CELLULAR PATHOLOGY, AS BASED UPON PHYSIOLOGICAL AND PATHOLOGICAL HISTOLOGY. With 144 Engravings on Wood. 8vo. cloth, 16*s.*

MR. H. T. CHAPMAN, F.R.C.S.

I.

THE TREATMENT OF OBSTINATE ULCERS AND CUTANEOUS ERUPTIONS OF THE LEG WITHOUT CONFINEMENT. Third Edition. Post 8vo. cloth, 3*s.* 6*d.*

II.

VARICOSE VEINS: their Nature, Consequences, and Treatment, Palliative and Curative. Second Edition. Post 8vo. cloth, 3*s.* 6*d.*

MR. PYE HENRY CHAVASSE, F.R.C.S.

I.

ADVICE TO A MOTHER ON THE MANAGEMENT OF HER OFFSPRING. Seventh Edition. Foolscap 8vo., 2*s.* 6*d.*

II.

ADVICE TO A WIFE ON THE MANAGEMENT OF HER OWN HEALTH. With an Introductory Chapter, especially addressed to a Young Wife. Sixth Edition. Fcap. 8vo., 2*s.* 6*d.*

MR. LE GROS CLARK, F.R.C.S.

OUTLINES OF SURGERY; being an Epitome of the Lectures on the Principles and the Practice of Surgery, delivered at St. Thomas's Hospital. Fcap. 8vo. cloth, 5*s.*

MR. JOHN CLAY, M.R.C.S.

KIWISCH ON DISEASES OF THE OVARIES: Translated, by permission, from the last German Edition of his Clinical Lectures on the Special Pathology and Treatment of the Diseases of Women. With Notes, and an Appendix on the Operation of Ovariotomy. Royal 12mo. cloth, 16*s.*

MR. COLLIS, M.B.DUB., F.R.C.S.I.

THE DIAGNOSIS AND TREATMENT OF CANCER AND THE TUMOURS ANALOGOUS TO IT. With coloured Plates. 8vo. cloth, 14*s.*

DR. CONOLLY.

THE CONSTRUCTION AND GOVERNMENT OF LUNATIC ASYLUMS AND HOSPITALS FOR THE INSANE. With Plans. Post 8vo. cloth, 6*s.*

MR. COOLEY.

COMPREHENSIVE SUPPLEMENT TO THE PHARMACOPŒIAS.

THE CYCLOPÆDIA OF PRACTICAL RECEIPTS, PROCESSES, AND COLLATERAL INFORMATION IN THE ARTS, MANUFACTURES, PROFESSIONS, AND TRADES, INCLUDING MEDICINE, PHARMACY, AND DOMESTIC ECONOMY; designed as a General Book of Reference for the Manufacturer, Tradesman, Amateur, and Heads of Families. Fourth and greatly enlarged Edition, 8vo. cloth, 28*s.*

MR. W. WHITE COOPER.

I.

ON WOUNDS AND INJURIES OF THE EYE. Illustrated by 17 Coloured Figures and 41 Woodcuts. 8vo. cloth, 12*s.*

II.

ON NEAR SIGHT, AGED SIGHT, IMPAIRED VISION, AND THE MEANS OF ASSISTING SIGHT. With 31 Illustrations on Wood. Second Edition. Fcap. 8vo. cloth, 7*s.* 6*d.*

SIR ASTLEY COOPER, BART., F.R.S.

ON THE STRUCTURE AND DISEASES OF THE TESTIS. With 24 Plates. Second Edition. Royal 4to., 20*s.*

MR. COOPER.

A DICTIONARY OF PRACTICAL SURGERY AND ENCYCLOPÆDIA OF SURGICAL SCIENCE. New Edition, brought down to the present time. By SAMUEL A. LANE, F.R.C.S., assisted by various eminent Surgeons. Vol. I., 8vo. cloth, £1. 5*s.*

MR. HOLMES COOTE, F.R.C.S.

A REPORT ON SOME IMPORTANT POINTS IN THE TREATMENT OF SYPHILIS. 8vo. cloth, 5*s.*

DR. COTTON.

I.

ON CONSUMPTION: Its Nature, Symptoms, and Treatment. To which Essay was awarded the Fothergillian Gold Medal of the Medical Society of London. Second Edition. 8vo. cloth, 8*s.*

II.

PHTHISIS AND THE STETHOSCOPE; OR, THE PHYSICAL SIGNS OF CONSUMPTION. Third Edition. Foolscap 8vo. cloth, 3*s.*

MR. COULSON.

I.

ON DISEASES OF THE BLADDER AND PROSTATE GLAND. New Edition, revised. *In Preparation.*

II.

ON LITHOTRITY AND LITHOTOMY; with Engravings on Wood. 8vo. cloth, 8*s.*

MR. WILLIAM CRAIG, L.F.P.S., GLASGOW.

ON THE INFLUENCE OF VARIATIONS OF ELECTRIC TENSION AS THE REMOTE CAUSE OF EPIDEMIC AND OTHER DISEASES. 8vo. cloth, 10*s.*

MR. CURLING, F.R.S.

I.

OBSERVATIONS ON DISEASES OF THE RECTUM. Third Edition. 8vo. cloth, 7*s.* 6*d.*

II.

A PRACTICAL TREATISE ON DISEASES OF THE TESTIS, SPERMATIC CORD, AND SCROTUM. Second Edition, with Additions. 8vo. cloth, 14*s.*

DR. DALRYMPLE, M.R.C.P., F.R.C.S.

THE CLIMATE OF EGYPT: METEOROLOGICAL AND MEDICAL OBSERVATIONS, with Practical Hints for Invalid Travellers. Post 8vo. cloth, 4*s.*

MR. JOHN DALRYMPLE, F.R.S., F.R.C.S.

PATHOLOGY OF THE HUMAN EYE. Complete in Nine Fasciculi: imperial 4to., 20*s.* each; half-bound morocco, gilt tops, 9*l.* 15*s.*

DR. HERBERT DAVIES.

ON THE PHYSICAL DIAGNOSIS OF DISEASES OF THE LUNGS AND HEART. Second Edition. Post 8vo. cloth, 8*s.*

DR. DAVEY.

I.

THE GANGLIONIC NERVOUS SYSTEM: its Structure, Functions, and Diseases. 8vo. cloth, 9*s.*

II.

ON THE NATURE AND PROXIMATE CAUSE OF INSANITY. Post 8vo. cloth, 3*s.*

MR. DIXON.

A GUIDE TO THE PRACTICAL STUDY OF DISEASES OF THE EYE. Second Edition. Post 8vo. cloth, 9*s.*

DR. DOBELL.

I.

DEMONSTRATIONS OF DISEASES IN THE CHEST, AND THEIR PHYSICAL DIAGNOSIS. With Coloured Plates. 8vo. cloth, 12*s.* 6*d.*

II.

LECTURES ON THE GERMS AND VESTIGES OF DISEASE, and on the Prevention of the Invasion and Fatality of Disease by Periodical Examinations. 8vo. cloth, 6*s.* 6*d.*

III.

A MANUAL OF DIET AND REGIMEN FOR PHYSICIAN AND PATIENT. Third Edition (for the year 1865). Crown 8vo. cloth, 1*s.* 6*d.*

DR. TOOGOOD DOWNING.

NEURALGIA: its various Forms, Pathology, and Treatment. The Jacksonian Prize Essay for 1850. 8vo. cloth, 10*s.* 6*d.*

DR. DRUITT, F.R.C.S.

THE SURGEON'S VADE-MECUM; with numerous Engravings on Wood. Eighth Edition. Foolscap 8vo. cloth, 12*s.* 6*d.*

MR. DUNN, F.R.C.S.

AN ESSAY ON PHYSIOLOGICAL PSYCHOLOGY. 8vo. cloth, 4*s.*

SIR JAMES EYRE, M.D.

I.

THE STOMACH AND ITS DIFFICULTIES. Fifth Edition. Fcap. 8vo. cloth, 2*s.* 6*d.*

II.

PRACTICAL REMARKS ON SOME EXHAUSTING DISEASES. Second Edition. Post 8vo. cloth, 4*s.* 6*d.*

DR. FENWICK.

ON SCROFULA AND CONSUMPTION. Clergyman's Sore Throat, Catarrh, Croup, Bronchitis, Asthma. Fcap. 8vo., 2*s.* 6*d.*

MR. FERGUSSON, F.R.S.

A SYSTEM OF PRACTICAL SURGERY; with numerous Illustrations on Wood. Fourth Edition. Fcap. 8vo. cloth, 12*s.* 6*d.*

MR. FLOWER, F.R.C.S.

DIAGRAMS OF THE NERVES OF THE HUMAN BODY, exhibiting their Origin, Divisions, and Connexions, with their Distribution to the various Regions of the Cutaneous Surface, and to all the Muscles. Folio, containing Six Plates, 14*s.*

MR. FOWNES, PH.D., F.R.S.

I.

A MANUAL OF CHEMISTRY; with 187 Illustrations on Wood. Ninth Edition. Fcap. 8vo. cloth, 12*s.* 6*d.*

Edited by H. BENCE JONES, M.D., F.R.S., and A. W. HOFMANN, PH.D., F.R.S.

II.

CHEMISTRY, AS EXEMPLIFYING THE WISDOM AND BENEFICENCE OF GOD. Second Edition. Fcap. 8vo. cloth, 4*s.* 6*d.*

III.

INTRODUCTION TO QUALITATIVE ANALYSIS. Post 8vo. cloth, 2*s.*

DR. D. J. T. FRANCIS.

CHANGE OF CLIMATE; considered as a Remedy in Dyspeptic, Pulmonary, and other Chronic Affections; with an Account of the most Eligible Places of Residence for Invalids, at different Seasons of the Year. Post 8vo. cloth, 8*s.* 6*d.*

DR. W. FRAZER.

ELEMENTS OF MATERIA MEDICA; containing the Chemistry and Natural History of Drugs—their Effects, Doses, and Adulterations. Second Edition. 8vo. cloth, 10*s.* 6*d.*

MR. J. G. FRENCH, F.R.C.S.

THE NATURE OF CHOLERA INVESTIGATED. Second Edition. 8vo. cloth, 4*s.*

C. REMIGIUS FRESENIUS.

A SYSTEM OF INSTRUCTION IN CHEMICAL ANALYSIS, Edited by LLOYD BULLOCK, F.C.S.

QUALITATIVE. Sixth Edition, with Coloured Plate illustrating Spectrum Analysis. 8vo. cloth, 10*s.* 6*d.*——QUANTITATIVE. Third Edition. 8vo. cloth, 16*s.*

DR. FULLER.

I.

ON DISEASES OF THE CHEST, including Diseases of the Heart and Great Vessels. With Engravings. 8vo. cloth, 12*s.* 6*d.*

II.

ON DISEASES OF THE HEART AND GREAT VESSELS. 8vo. cloth, 7*s.* 6*d.*

III.

ON RHEUMATISM, RHEUMATIC GOUT, AND SCIATICA: their Pathology, Symptoms, and Treatment. Third Edition. 8vo. cloth, 12*s.* 6*d.*

DR. GAIRDNER.

ON GOUT; its History, its Causes, and its Cure. Fourth Edition. Post 8vo. cloth, 8*s.* 6*d.*

MR. GALLOWAY.

I.

THE FIRST STEP IN CHEMISTRY. Third Edition. Fcap. 8vo. cloth, 5*s.*

II.

THE SECOND STEP IN CHEMISTRY; or, the Student's Guide to the Higher Branches of the Science. With Engravings. 8vo. cloth, 10*s.*

III.

A MANUAL OF QUALITATIVE ANALYSIS. Fourth Edition. Post 8vo. cloth, 6*s.* 6*d.*

IV.

CHEMICAL TABLES. On Five Large Sheets, for School and Lecture Rooms. Second Edition. 4*s.* 6*d.*

MR. F. J. GANT, F.R.C.S.

I.

THE PRINCIPLES OF SURGERY: Clinical, Medical, and Operative. With Engravings. 8vo. cloth, 18*s.*

II.

THE IRRITABLE BLADDER: its Causes and Curative Treatment. Post 8vo. cloth, 4*s.* 6*d.*

DR. GIBB. M.R.C.P.

ON DISEASES OF THE THROAT AND WINDPIPE, as reflected by the Laryngoscope. Second Edition. With 116 Engravings. Post 8vo. cloth, 10*s.* 6*d.*

MRS. GODFREY.

ON THE NATURE, PREVENTION, TREATMENT, AND CURE OF SPINAL CURVATURES and DEFORMITIES of the CHEST and LIMBS, without ARTIFICIAL SUPPORTS or any MECHANICAL APPLIANCES. Third Edition, Revised and Enlarged. 8vo. cloth, 5*s.*

DR. GORDON, M.D., C.B.

CHINA, FROM A MEDICAL POINT OF VIEW, IN 1860 AND 1861; With a Chapter on Nagasaki as a Sanatarium. With Plans. 8vo. cloth, 10*s.* 6*d.*

DR. GRANVILLE, F.R.S.

I.

THE MINERAL SPRINGS OF VICHY: their Efficacy in the Treatment of Gout, Indigestion, Gravel, &c. 8vo. cloth, 3*s.*

II.

ON SUDDEN DEATH. Post 8vo., 2*s.* 6*d.*

DR. GRAVES, M.D., F.R.S.

STUDIES IN PHYSIOLOGY AND MEDICINE. Edited by Dr. Stokes. With Portrait and Memoir. 8vo. cloth, 14*s.*

MR. GRIFFITHS.

CHEMISTRY OF THE FOUR SEASONS—Spring, Summer, Autumn, Winter. Illustrated with Engravings on Wood. Second Edition. Foolscap 8vo. cloth, 7*s.* 6*d.*

DR. GULLY.

THE SIMPLE TREATMENT OF DISEASE; deduced from the Methods of Expectancy and Revulsion. 18mo. cloth, 4*s.*

DR. GUY AND DR. JOHN HARLEY.

HOOPER'S PHYSICIAN'S VADE-MECUM; OR, MANUAL OF THE PRINCIPLES AND PRACTICE OF PHYSIC. Seventh Edition, considerably enlarged, and rewritten. Foolscap 8vo. cloth, 12*s.* 6*d.*

GUY'S HOSPITAL REPORTS. Third Series. Vols. I. to X., 8vo., 7*s.* 6*d.* each.

DR. HABERSHON, F.R.C.P.

I.

PATHOLOGICAL AND PRACTICAL OBSERVATIONS ON DISEASES OF THE ABDOMEN, comprising those of the Stomach and other Parts of the Alimentary Canal, Œsophagus, Stomach, Cæcum, Intestines, and Peritoneum. Second Edition, with Plates. 8vo. cloth, 14*s.*

II.

ON THE INJURIOUS EFFECTS OF MERCURY IN THE TREATMENT OF DISEASE. Post 8vo. cloth, 3*s.* 6*d.*

DR. C. RADCLYFFE HALL.

TORQUAY IN ITS MEDICAL ASPECT AS A RESORT FOR PULMONARY INVALIDS. Post 8vo. cloth, 5*s.*

DR. MARSHALL HALL, F.R.S.

I.

PRONE AND POSTURAL RESPIRATION IN DROWNING AND OTHER FORMS OF APNŒA OR SUSPENDED RESPIRATION. Post 8vo. cloth. 5*s.*

II.

PRACTICAL OBSERVATIONS AND SUGGESTIONS IN MEDICINE. Second Series. Post 8vo. cloth, 8*s.* 6*d.*

MR. HARDWICH.

A MANUAL OF PHOTOGRAPHIC CHEMISTRY. With Engravings. Seventh Edition. Foolscap 8vo. cloth, 7*s.* 6*d.*

DR. J. BOWER HARRISON, M.D., M.R.C.P.

I.

LETTERS TO A YOUNG PRACTITIONER ON THE DISEASES OF CHILDREN. Foolscap 8vo. cloth, 3*s.*

II.

ON THE CONTAMINATION OF WATER BY THE POISON OF LEAD, and its Effects on the Human Body. Foolscap 8vo. cloth, 3*s.* 6*d.*

DR. HARTWIG.

I.

ON SEA BATHING AND SEA AIR. Second Edition. Fcap. 8vo., 2*s.* 6*d.*

II.

ON THE PHYSICAL EDUCATION OF CHILDREN. Fcap. 8vo., 2*s.* 6*d.*

DR. A. H. HASSALL.

I.

THE URINE, IN HEALTH AND DISEASE; being an Explanation of the Composition of the Urine, and of the Pathology and Treatment of Urinary and Renal Disorders. Second Edition. With 79 Engravings (23 Coloured). Post 8vo. cloth, 12*s.* 6*d.*

II.

THE MICROSCOPIC ANATOMY OF THE HUMAN BODY, IN HEALTH AND DISEASE. Illustrated with Several Hundred Drawings in Colour. Two vols. 8vo. cloth, £1. 10*s.*

MR. ALFRED HAVILAND, M.R.C.S.

CLIMATE, WEATHER, AND DISEASE; being a Sketch of the Opinions of the most celebrated Ancient and Modern Writers with regard to the Influence of Climate and Weather in producing Disease. With Four coloured Engravings. 8vo. cloth, 7*s.*

DR. HEADLAND.

ON THE ACTION OF MEDICINES IN THE SYSTEM. Being the Prize Essay to which the Medical Society of London awarded the Fothergillian Gold Medal for 1852. Third Edition. 8vo. cloth, 12*s.* 6*d.*

DR. HEALE.

I.

A TREATISE ON THE PHYSIOLOGICAL ANATOMY OF THE LUNGS. With Engravings. 8vo. cloth, 8*s.*

II.

A TREATISE ON VITAL CAUSES. 8vo. cloth, 9*s.*

MR. CHRISTOPHE HEATH, F.R.C.S.

I.

PRACTICAL ANATOMY: a Manual of Dissections. With numerous Engravings. Fcap. 8vo. cloth, 10*s.* 6*d.*

II.

A MANUAL OF MINOR SURGERY AND BANDAGING, FOR THE USE OF HOUSE-SURGEONS, DRESSERS, AND JUNIOR PRACTITIONERS. With Illustrations. Second Edition. Fcap. 8vo. cloth, 5*s.*

MR. HIGGINBOTTOM, F.R.S., F.R.C.S.E.

ON THE NITRATE OF SILVER: WITH FULL DIRECTIONS FOR ITS APPLICATION IN THE TREATMENT OF INFLAMMATION, WOUNDS, AND ULCERS. Part I., Second Edition, 5*s.*; Part II., 2*s.* 6*d.*

DR. HINDS.

THE HARMONIES OF PHYSICAL SCIENCE IN RELATION TO THE HIGHER SENTIMENTS; with Observations on Medical Studies, and on the Moral and Scientific Relations of Medical Life. Post 8vo. cloth, 4*s.*

MR. J. A. HINGESTON, M.R.C.S.

TOPICS OF THE DAY, MEDICAL, SOCIAL, AND SCIENTIFIC. Crown 8vo. cloth, 7*s.* 6*d.*

DR. HODGES.

THE NATURE, PATHOLOGY, AND TREATMENT OF PUERPERAL CONVULSIONS. Crown 8vo. cloth, 3*s.*

DR. DECIMUS HODGSON.

THE PROSTATE GLAND, AND ITS ENLARGEMENT IN OLD AGE. With 12 Plates. Royal 8vo. cloth, 6*s.*

MR. JABEZ HOGG.

A MANUAL OF OPHTHALMOSCOPIC SURGERY; being a Practical Treatise on the Use of the Ophthalmoscope in Diseases of the Eye. Third Edition. With Coloured Plates. 8vo. cloth, 10*s.* 6*d.*

MR. LUTHER HOLDEN, F.R.C.S.

I.

HUMAN OSTEOLOGY: with Plates, showing the Attachments of the Muscles. Third Edition. 8vo. cloth, 16*s.*

II.

A MANUAL OF THE DISSECTION OF THE HUMAN BODY. With Engravings on Wood. Second Edition. 8vo. cloth, 16*s.*

MR BARNARD HOLT, F.R.C.S.

ON THE IMMEDIATE TREATMENT OF STRICTURE OF THE URETHRA. Second Edition, Enlarged. 8vo. cloth, 3*s.*

MR. C. HOLTHOUSE.

I.

ON SQUINTING, PARALYTIC AFFECTIONS OF THE EYE, and CERTAIN FORMS OF IMPAIRED VISION. Fcap. 8vo. cloth, 4*s.* 6*d.*

II.

LECTURES ON STRABISMUS, delivered at the Westminster Hospital. 8vo. cloth, 4*s.*

DR. W. CHARLES HOOD.

SUGGESTIONS FOR THE FUTURE PROVISION OF CRIMINAL LUNATICS. 8vo. cloth, 5*s*. 6*d*.

DR. P. HOOD.

THE SUCCESSFUL TREATMENT OF SCARLET FEVER; also, OBSERVATIONS ON THE PATHOLOGY AND TREATMENT OF CROWING INSPIRATIONS OF INFANTS. Post 8vo. cloth, 5*s*.

MR. JOHN HORSLEY.

A CATECHISM OF CHEMICAL PHILOSOPHY; being a Familiar Exposition of the Principles of Chemistry and Physics. With Engravings on Wood. Designed for the Use of Schools and Private Teachers. Post 8vo. cloth, 6*s*. 6*d*.

MR. LUKE HOWARD, F.R.S.

ESSAY ON THE MODIFICATIONS OF CLOUDS. Third Edition, by W. D. and E. HOWARD. With 6 Lithographic Plates, from Pictures by Kenyon. 4to. cloth, 10*s*. 6*d*.

DR. HUFELAND.

THE ART OF PROLONGING LIFE. Second Edition. Edited by ERASMUS WILSON, F.R.S. Foolscap 8vo., 2*s*. 6*d*.

MR. W. CURTIS HUGMAN, F.R.C.S.

ON HIP-JOINT DISEASE; with reference especially to Treatment by Mechanical Means for the Relief of Contraction and Deformity of the Affected Limb. 8vo. cloth, 3*s*. 6*d*.

MR. HULKE, F.R.C.S.

A PRACTICAL TREATISE ON THE USE OF THE OPHTHALMOSCOPE. Being the Jacksonian Prize Essay for 1859. Royal 8vo. cloth, 8*s*.

DR. HENRY HUNT.

ON HEARTBURN AND INDIGESTION. 8vo. cloth, 5*s*.

PROFESSOR HUXLEY, F.R.S.

LECTURES ON THE ELEMENTS OF COMPARATIVE ANATOMY.—ON CLASSIFICATON AND THE SKULL. With 111 Illustrations. 8vo. cloth, 10*s*. 6*d*.

MR. JONATHAN HUTCHINSON, F.R.C.S.

A CLINICAL MEMOIR ON CERTAIN DISEASES OF THE EYE AND EAR, CONSEQUENT ON INHERITED SYPHILIS; with an appended Chapter of Commentaries on the Transmission of Syphilis from Parent to Offspring, and its more remote Consequences. With Plates and Woodcuts, 8vo. cloth, 9*s*.

DR. INMAN, M.R.C.P.

I.

ON MYALGIA: ITS NATURE, CAUSES, AND TREATMENT; being a Treatise on Painful and other Affections of the Muscular System. Second Edition. 8vo. cloth, 9*s*.

II.

FOUNDATION FOR A NEW THEORY AND PRACTICE OF MEDICINE. Second Edition. Crown 8vo. cloth, 10*s*.

DR. ARTHUR JACOB, F.R.C.S.

A TREATISE ON THE INFLAMMATIONS OF THE EYE-BALL. Foolscap 8vo. cloth, 5*s*.

DR. JAGO, M.D.OXON., A.B.CANTAB.

ENTOPTICS, WITH ITS USES IN PHYSIOLOGY AND MEDICINE. With 54 Engravings. Crown 8vo. cloth, 5*s.*

MR. J. H. JAMES, F.R.C.S.

PRACTICAL OBSERVATIONS ON THE OPERATIONS FOR STRANGULATED HERNIA. 8vo. cloth, 5*s.*

DR. PROSSER JAMES, M.D.

SORE-THROAT: ITS NATURE, VARIETIES, AND TREATMENT; including the Use of the LARYNGOSCOPE as an Aid to Diagnosis. Post 8vo. cloth, 4*s.* 6*d.*

DR. HANDFIELD JONES, M.B., F.R.C.P.

CLINICAL OBSERVATIONS ON FUNCTIONAL NERVOUS DISORDERS. Post 8vo. cloth, 10*s.* 6*d.*

DR. HANDFIELD JONES, F.R.S., & DR. EDWARD H. SIEVEKING.

A MANUAL OF PATHOLOGICAL ANATOMY. Illustrated with numerous Engravings on Wood. Foolscap 8vo. cloth, 12*s.* 6*d.*

DR. JAMES JONES, M.D., M.R.C.P.

ON THE USE OF PERCHLORIDE OF IRON AND OTHER CHALYBEATE SALTS IN THE TREATMENT OF CONSUMPTION. Crown 8vo. cloth, 3*s.* 6*d.*

MR. WHARTON JONES, F.R.S.

I.

A MANUAL OF THE PRINCIPLES AND PRACTICE OF OPHTHALMIC MEDICINE AND SURGERY; illustrated with Engravings, plain and coloured. Second Edition. Foolscap 8vo. cloth, 12*s.* 6*d.*

II.

THE WISDOM AND BENEFICENCE OF THE ALMIGHTY, AS DISPLAYED IN THE SENSE OF VISION; being the Actonian Prize Essay for 1851. With Illustrations on Steel and Wood. Foolscap 8vo. cloth, 4*s.* 6*d.*

III.

DEFECTS OF SIGHT: their Nature, Causes, Prevention, and General Management. Fcap. 8vo. 2*s.* 6*d.*

IV.

A CATECHISM OF THE MEDICINE AND SURGERY OF THE EYE AND EAR. For the Clinical Use of Hospital Students. Fcap. 8vo. 2*s.* 6*d.*

V.

A CATECHISM OF THE PHYSIOLOGY AND PHILOSOPHY OF BODY, SENSE, AND MIND. For Use in Schools and Colleges. Fcap. 8vo., 2*s.* 6*d.*

MR. FURNEAUX JORDAN, M.R.C.S.

AN INTRODUCTION TO CLINICAL SURGERY; WITH A Method of Investigating and Reporting Surgical Cases. Fcap. 8vo. cloth, 5*s.*

MR. JUDD.

A PRACTICAL TREATISE ON URETHRITIS AND SYPHILIS: including Observations on the Power of the Menstruous Fluid, and of the Discharge from Leucorrhœa and Sores to produce Urethritis: with a variety of Examples, Experiments, Remedies, and Cures. 8vo. cloth, £1. 5*s.*

DR. LAENNEC.

A MANUAL OF AUSCULTATION AND PERCUSSION. Translated and Edited by J. B. SHARPE, M.R.C.S. 3*s*.

DR. LANE, M.A.

HYDROPATHY; OR, HYGIENIC MEDICINE. An Explanatory Essay. Second Edition. Post 8vo. cloth, 5*s*.

MR. LAWRENCE, F.R.S.

I.

LECTURES ON SURGERY. 8vo. cloth, 16*s*.

II.

A TREATISE ON RUPTURES. The Fifth Edition, considerably enlarged. 8vo. cloth, 16*s*.

DR. LEARED, M.R.C.P.

IMPERFECT DIGESTION: ITS CAUSES AND TREATMENT. Third Edition. Foolscap 8vo. cloth, 4*s*.

DR. EDWIN LEE.

I.

THE EFFECT OF CLIMATE ON TUBERCULOUS DISEASE, with Notices of the chief Foreign Places of Winter Resort. Small 8vo. cloth, 4*s*. 6*d*.

II.

THE WATERING PLACES OF ENGLAND, CONSIDERED with Reference to their Medical Topography. Fourth Edition. Fcap. 8vo. cloth, 7*s*. 6*d*.

III.

THE BATHS OF GERMANY. Fourth Edition. Post 8vo. cloth, 7*s*.

IV.

HOMŒOPATHY AND HYDROPATHY IMPARTIALLY APPRECIATED. With Notes illustrative of the Influence of the Mind over the Body. Fourth Edition. Post 8vo. cloth, 3*s*. 6*d*.

MR. HENRY LEE, F.R.C.S.

ON SYPHILIS. Second Edition. With Coloured Plates. 8vo. cloth, 10*s*.

DR. ROBERT LEE, F.R.S.

I.

CONSULTATIONS IN MIDWIFERY. Foolscap 8vo. cloth, 4*s*. 6*d*.

II.

A TREATISE ON THE SPECULUM; with Three Hundred Cases. 8vo. cloth, 4*s*. 6*d*.

III.

CLINICAL REPORTS OF OVARIAN AND UTERINE DISEASES, with Commentaries. Foolscap 8vo. cloth, 6*s*. 6*d*.

IV.

CLINICAL MIDWIFERY: comprising the Histories of 545 Cases of Difficult, Preternatural, and Complicated Labour, with Commentaries. Second Edition. Foolscap 8vo. cloth, 5*s*.

V.

PRACTICAL OBSERVATIONS ON DISEASES OF THE UTERUS. With coloured Plates. Two Parts. Imperial 4to., 7*s*. 6*d*. each Part.

DR. LEISHMAN, M.D., F.F.P.S.

THE MECHANISM OF PARTURITION: An Essay, Historical and Critical. With Engravings. 8vo. cloth, 5*s*.

MR. LISTON, F.R.S.

PRACTICAL SURGERY. Fourth Edition. 8vo. cloth, 22*s*.

MR. H. W. LOBB, L.S.A., M.R.C.S.E.

ON SOME OF THE MORE OBSCURE FORMS OF NERVOUS AFFECTIONS, THEIR PATHOLOGY AND TREATMENT. Re-issue, with the Chapter on Galvanism entirely Re-written. With Engravings. 8vo. cloth, 8*s.*

DR. LOGAN, M.D., M.R.C.P.LOND.

ON OBSTINATE DISEASES OF THE SKIN. Foolscap 8vo. cloth, 2*s.* 6*d.*

LONDON HOSPITAL.

CLINICAL LECTURES AND REPORTS BY THE MEDICAL AND SURGICAL STAFF. With Illustrations. Vol. I. 8vo. cloth, 7*s.* 6*d.*

LONDON MEDICAL SOCIETY OF OBSERVATION.

WHAT TO OBSERVE AT THE BED-SIDE, AND AFTER DEATH. Published by Authority. Second Edition. Foolscap 8vo. cloth, 4*s.* 6*d.*

DR. MACKENZIE, M.D., M.R.C.P.

THE PATHOLOGY AND TREATMENT OF PHLEGMASIA DOLENS, as deduced from Clinical and Physiological Researches. Lettsomian Lectures on Midwifery. 8vo. cloth, 6*s.*

MR. M'CLELLAND, F.L.S., F.G.S.

THE MEDICAL TOPOGRAPHY, OR CLIMATE AND SOILS, OF BENGAL AND THE N. W. PROVINCES. Post 8vo. cloth, 4*s.* 6*d.*

DR. MACLACHLAN, M.D., F.R.C.P.L.

THE DISEASES AND INFIRMITIES OF ADVANCED LIFE. 8vo. cloth, 16*s.*

DR. GEORGE H. B. MACLEOD, F.R.C.S.E.

I.

OUTLINES OF SURGICAL DIAGNOSIS. 8vo. cloth, 12*s.* 6*d.*

II.

NOTES ON THE SURGERY OF THE CRIMEAN WAR; with REMARKS on GUN-SHOT WOUNDS. 8vo. cloth, 10*s.* 6*d.*

MR. JOSEPH MACLISE, F.R.C.S.

I.

SURGICAL ANATOMY. A Series of Dissections, illustrating the Principal Regions of the Human Body.
The Second Edition, imperial folio, cloth, £3. 12*s.*; half-morocco, £4. 4*s.*

II.

ON DISLOCATIONS AND FRACTURES. This Work is Uniform with the Author's "Surgical Anatomy;" each Fasciculus contains Four beautifully executed Lithographic Drawings. Imperial folio, cloth, £2. 10*s.*; half-morocco, £2. 17*s.*

DR. MCNICOLL, M.R.C.P.

A HAND-BOOK FOR SOUTHPORT, MEDICAL & GENERAL; with Copious Notices of the Natural History of the District. Second Edition. Post 8vo. cloth, 3*s.* 6*d.*

DR. MARCET, F.R.S.

I.

ON THE COMPOSITION OF FOOD, AND HOW IT IS ADULTERATED; with Practical Directions for its Analysis. 8vo. cloth, 6*s.* 6*d.*

II.

ON CHRONIC ALCOHOLIC INTOXICATION; with an INQUIRY INTO THE INFLUENCE OF THE ABUSE OF ALCOHOL AS A PREDISPOSING CAUSE OF DISEASE. Second Edition, much enlarged. Foolscap 8vo. cloth, 4*s.* 6*d.*

DR. MARKHAM.

I.

DISEASES OF THE HEART: THEIR PATHOLOGY, DIAGNOSIS, AND TREATMENT. Second Edition. Post 8vo. cloth, 6*s.*

II.

SKODA ON AUSCULTATION AND PERCUSSION. Post 8vo. cloth, 6*s.*

SIR RANALD MARTIN, K.C.B., F.R.S.

INFLUENCE OF TROPICAL CLIMATES IN PRODUCING THE ACUTE ENDEMIC DISEASES OF EUROPEANS; including Practical Observations on their Chronic Sequelæ under the Influences of the Climate of Europe. Second Edition, much enlarged. 8vo. cloth, 20*s.*

DR. MASSY.

ON THE EXAMINATION OF RECRUITS; intended for the Use of Young Medical Officers on Entering the Army. 8vo. cloth, 5*s.*

MR. C. F. MAUNDER, F.R.C.S.

OPERATIVE SURGERY. With 158 Engravings. Post 8vo. 6*s.*

DR. MAYNE.

I.

AN EXPOSITORY LEXICON OF THE TERMS, ANCIENT AND MODERN, IN MEDICAL AND GENERAL SCIENCE, including a complete MEDICAL AND MEDICO-LEGAL VOCABULARY. Complete in 10 Parts, price 5*s.* each. The entire work, cloth, £2. 10*s.*

II.

A MEDICAL VOCABULARY; or, an Explanation of all Names, Synonymes, Terms, and Phrases used in Medicine and the relative branches of Medical Science, intended specially as a Book of Reference for the Young Student. Second Edition. Fcap. 8vo. cloth, 8*s.* 6*d.*

DR. MERYON, M.D., F.R.C.P.

PATHOLOGICAL AND PRACTICAL RESEARCHES ON THE VARIOUS FORMS OF PARALYSIS. 8vo. cloth, 6*s.*

DR. MILLINGEN.

ON THE TREATMENT AND MANAGEMENT OF THE INSANE; with Considerations on Public and Private Lunatic Asylums. 18mo. cloth, 4*s.* 6*d.*

DR. W. J. MOORE, M.D.

I.

HEALTH IN THE TROPICS; or, Sanitary Art applied to Europeans in India. 8vo. cloth, 9*s.*

II.

A MANUAL OF THE DISEASES OF INDIA. Fcap. 8vo. cloth, 5*s.*

PROFESSOR MULDER, UTRECHT.

THE CHEMISTRY OF WINE. Edited by H. Bence Jones, M.D., F.R.S. Fcap. 8vo. cloth, 6*s.*

DR. BIRKBECK NEVINS.

THE PRESCRIBER'S ANALYSIS OF THE BRITISH PHARMACOPEIA. Second Edition, enlarged to 264 pp. 32mo. cloth, 3*s.* 6*d.*

DR. NOBLE.

THE HUMAN MIND IN ITS RELATIONS WITH THE BRAIN AND NERVOUS SYSTEM. Post 8vo. cloth, 4*s.* 6*d.*

MR. NUNNELEY, F.R.C.S.E.

I.

ON THE ORGANS OF VISION: THEIR ANATOMY AND PHYSIOLOGY. With Plates, 8vo. cloth, 15*s.*

II.

A TREATISE ON THE NATURE, CAUSES, AND TREATMENT OF ERYSIPELAS. 8vo. cloth, 10*s.* 6*d.*

DR. O'REILLY.

THE PLACENTA, THE ORGANIC NERVOUS SYSTEM, THE BLOOD, THE OXYGEN, AND THE ANIMAL NERVOUS SYSTEM, PHYSIOLOGICALLY EXAMINED. With Engravings. 8vo. cloth, 5*s.*

MR. LANGSTON PARKER.

THE MODERN TREATMENT OF SYPHILITIC DISEASES, both Primary and Secondary; comprising the Treatment of Constitutional and Confirmed Syphilis, by a safe and successful Method. Fourth Edition, 8vo. cloth, 10*s.*

DR. PARKES, F.R.C.P.

I.

A MANUAL OF PRACTICAL HYGIENE; intended especially for the Medical Officers of the Army. With Plates and Woodcuts. 8vo. cloth, 16*s.*

II.

THE URINE: ITS COMPOSITION IN HEALTH AND DISEASE, AND UNDER THE ACTION OF REMEDIES. 8vo. cloth, 12*s.*

DR. PARKIN, M.D., F.R.C.S.

THE CAUSATION AND PREVENTION OF DISEASE; with the Laws regulating the Extrication of Malaria from the Surface, and its Diffusion in the surrounding Air. 8vo. cloth, 5*s.*

MR. JAMES PART, F.R.C.S.

THE MEDICAL AND SURGICAL POCKET CASE BOOK, for the Registration of important Cases in Private Practice, and to assist the Student of Hospital Practice. Second Edition. 2*s.* 6*d.*

DR. PAVY, M.D., F.R.S., F.R.C.P.

DIABETES: RESEARCHES ON ITS NATURE AND TREATMENT. 8vo. cloth, 8*s.* 6*d.*

DR. THOMAS B. PEACOCK, M.D.

ON THE INFLUENZA, OR EPIDEMIC CATARRHAL FEVER OF 1847-8. 8vo. cloth, 5*s.* 6*d.*

DR. PEET, M.D., F.R.C.P.

THE PRINCIPLES AND PRACTICE OF MEDICINE; Designed chiefly for Students of Indian Medical Colleges. 8vo. cloth, 16*s.*

DR. PEREIRA, F.R.S.

SELECTA E PRÆSCRIPTIS. Fourteenth Edition. 24mo. cloth, 5*s.*

DR. PICKFORD.

HYGIENE; or, Health as Depending upon the Conditions of the Atmosphere, Food and Drinks, Motion and Rest, Sleep and Wakefulness, Secretions, Excretions, and Retentions, Mental Emotions, Clothing, Bathing, &c. Vol. I. 8vo. cloth, 9*s.*

MR. PIRRIE, F.R.S.E.

THE PRINCIPLES AND PRACTICE OF SURGERY. With numerous Engravings on Wood. Second Edition. 8vo. cloth, 24*s.*

PHARMACOPŒIA COLLEGII REGALIS MEDICORUM LONDINENSIS. 8vo. cloth, 9*s.*; or 24mo. 5*s.*

IMPRIMATUR.

Hic liber, cui titulus, PHARMACOPŒIA COLLEGII REGALIS MEDICORUM LONDINENSIS. Datum ex Ædibus Collegii in comitiis censoriis, Novembris Mensis 14to 1850.

JOHANNES AYRTON PARIS. *Præses.*

PROFESSORS PLATTNER & MUSPRATT.

THE USE OF THE BLOWPIPE IN THE EXAMINATION OF MINERALS, ORES, AND OTHER METALLIC COMBINATIONS. Illustrated by numerous Engravings on Wood. Third Edition. 8vo. cloth, 10*s.* 6*d.*

DR. HENRY F. A. PRATT, M.D., M.R.C.P.

I.

THE GENEALOGY OF CREATION, newly Translated from the Unpointed Hebrew Text of the Book of Genesis, showing the General Scientific Accuracy of the Cosmogony of Moses and the Philosophy of Creation. 8vo. cloth, 14*s.*

II.

ON ECCENTRIC AND CENTRIC FORCE: A New Theory of Projection. With Engravings. 8vo. cloth, 10*s.*

III.

ON ORBITAL MOTION: The Outlines of a System of Physical Astronomy. With Diagrams. 8vo. cloth, 7*s.* 6*d.*

THE PRESCRIBER'S PHARMACOPŒIA; containing all the Medicines in the British Pharmacopœia, arranged in Classes according to their Action, with their Composition and Doses. By a Practising Physician. Fifth Edition. 32mo. cloth, 2*s.* 6*d.*; roan tuck (for the pocket), 3*s.* 6*d.*

DR. JOHN ROWLISON PRETTY.

AIDS DURING LABOUR, including the Administration of Chloroform, the Management of Placenta and Post-partum Hæmorrhage. Fcap. 8vo. cloth, 4*s.* 6*d.*

MR. LAKE PRICE.

PHOTOGRAPHIC MANIPULATION: Treating of the Practice of the Art, and its various appliances to Nature. With Fifty Engravings on Wood. Post 8vo. cloth, 6*s.* 6*d.*

DR. PRIESTLEY.

LECTURES ON THE DEVELOPMENT OF THE GRAVID UTERUS. 8vo. cloth, 5*s.* 6*d.*

DR. RADCLIFFE, F.R.C.P.L.

LECTURES ON EPILEPSY, PAIN, PARALYSIS, AND CERTAIN OTHER DISORDERS OF THE NERVOUS SYSTEM, delivered at the Royal College of Physicians in London. Post 8vo. cloth, 7*s.* 6*d.*

MR. RAINEY.

ON THE MODE OF FORMATION OF SHELLS OF ANIMALS,
OF BONE, AND OF SEVERAL OTHER STRUCTURES, by a Process of Molecular Coalescence, Demonstrable in certain Artificially-formed Products. Fcap. 8vo. cloth, 4*s.* 6*d.*

DR. F. H. RAMSBOTHAM.

THE PRINCIPLES AND PRACTICE OF OBSTETRIC MEDICINE AND SURGERY.
Illustrated with One Hundred and Twenty Plates on Steel and Wood; forming one thick handsome volume. Fourth Edition. 8vo. cloth, 22*s.*

DR. RAMSBOTHAM.

PRACTICAL OBSERVATIONS ON MIDWIFERY,
with a Selection of Cases. Second Edition. 8vo. cloth, 12*s.*

PROFESSOR REDWOOD, PH. D.

A SUPPLEMENT TO THE PHARMACOPŒIA:
A concise but comprehensive Dispensatory, and Manual of Facts and Formulæ, for the use of Practitioners in Medicine and Pharmacy. Third Edition. 8vo. cloth, 22*s.*

DR. DU BOIS REYMOND.

ANIMAL ELECTRICITY;
Edited by H. BENCE JONES, M.D., F.R.S. With Fifty Engravings on Wood. Foolscap 8vo. cloth, 6*s.*

DR. REYNOLDS, M.D.LOND.

I.

EPILEPSY: ITS SYMPTOMS, TREATMENT, AND RELATION TO OTHER CHRONIC CONVULSIVE DISEASES.
8vo. cloth, 10*s.*

II.

THE DIAGNOSIS OF DISEASES OF THE BRAIN, SPINAL CORD, AND THEIR APPENDAGES.
8vo. cloth, 8*s.*

DR. B. W. RICHARDSON.

I.

ON THE CAUSE OF THE COAGULATION OF THE BLOOD.
Being the ASTLEY COOPER PRIZE ESSAY for 1856. With a Practical Appendix. 8vo. cloth, 16*s.*

II.

THE HYGIENIC TREATMENT OF PULMONARY CONSUMPTION.
8vo. cloth, 5*s.* 6*d.*

III.

THE ASCLEPIAD.
Vol. I., Clinical Essays. 8vo. cloth, 6*s.* 6*d.*

MR. WILLIAM ROBERTS.

AN ESSAY ON WASTING PALSY;
being a Systematic Treatise on the Disease hitherto described as ATROPHIE MUSCULAIRE PROGRESSIVE. With Four Plates. 8vo. cloth, 7*s.* 6*d.*

DR. ROUTH.

INFANT FEEDING, AND ITS INFLUENCE ON LIFE;
Or, the Causes and Prevention of Infant Mortality. Second Edition. Fcap. 8vo. cloth, 6*s.*

DR. W. H. ROBERTSON.

I.

THE NATURE AND TREATMENT OF GOUT.

8vo. cloth, 10*s*. 6*d*.

II.

A TREATISE ON DIET AND REGIMEN.

Fourth Edition. 2 vols. post 8vo. cloth, 12*s*.

DR. ROWE.

NERVOUS DISEASES, LIVER AND STOMACH COMPLAINTS, LOW SPIRITS, INDIGESTION, GOUT, ASTHMA, AND DISORDERS PRODUCED BY TROPICAL CLIMATES. With Cases. Sixteenth Edition. Fcap. 8vo. 2*s*. 6*d*.

DR. ROYLE, F.R.S., AND DR. HEADLAND, M.D.

A MANUAL OF MATERIA MEDICA AND THERAPEUTICS. With numerous Engravings on Wood. Fourth Edition. Fcap. 8vo. cloth, 12*s*. 6*d*.

MR. RUMSEY, F.R.C.S.

ESSAYS ON STATE MEDICINE. 8vo. cloth, 10*s*. 6*d*.

DR. RYAN, M.D.

INFANTICIDE: ITS LAW, PREVALENCE, PREVENTION, AND HISTORY. 8vo. cloth, 5*s*.

ST. BARTHOLOMEW'S HOSPITAL.

A DESCRIPTIVE CATALOGUE OF THE ANATOMICAL MUSEUM. Vol. I. (1846), Vol. II. (1851), Vol. III. (1862), 8vo. cloth, 5*s*. each.

DR. SALTER, F.R.S.

ON ASTHMA: its Pathology, Causes, Consequences, and Treatment. 8vo. cloth, 10*s*.

DR. SAVAGE, M.D.LOND., F.R.C.S.

THE SURGERY OF THE FEMALE PELVIC ORGANS, in a Series of Plates taken from Nature, with Physiological and Pathological References. Royal 4to. cloth, 20*s*.

*** These Plates give 40 Illustrations taken from original Dissections, and are drawn and coloured in the highest degree of art.

MR. SAVORY.

A COMPENDIUM OF DOMESTIC MEDICINE, AND COMPANION TO THE MEDICINE CHEST; intended as a Source of Easy Reference for Clergymen, and for Families residing at a Distance from Professional Assistance. Sixth Edition. 12mo. cloth, 5*s*.

DR. SCHACHT.

THE MICROSCOPE, AND ITS APPLICATION TO VEGETABLE ANATOMY AND PHYSIOLOGY. Edited by FREDERICK CURREY, M.A. Fcap. 8vo. cloth, 6*s*.

DR. SCORESBY-JACKSON, M.D., F.R.S.E.

MEDICAL CLIMATOLOGY; or, a Topographical and Meteorological Description of the Localities resorted to in Winter and Summer by Invalids of various classes both at Home and Abroad. With an Isothermal Chart. Post 8vo. cloth, 12*s*.

DR. SEMPLE.

ON COUGH: its Causes, Varieties, and Treatment. With some practical Remarks on the Use of the Stethoscope as an aid to Diagnosis. Post 8vo. cloth, 4*s*. 6*d*.

DR. SEYMOUR.

I.

ILLUSTRATIONS OF SOME OF THE PRINCIPAL DISEASES OF THE OVARIA: their Symptoms and Treatment; to which are prefixed Observations on the Structure and Functions of those parts in the Human Being and in Animals. With 14 folio plates, 12*s.*

II.

THE NATURE AND TREATMENT OF DROPSY; considered especially in reference to the Diseases of the Internal Organs of the Body, which most commonly produce it. 8vo. 5*s.*

DR. SHAPTER, M.D., F.R.C.P.

THE CLIMATE OF THE SOUTH OF DEVON, AND ITS INFLUENCE UPON HEALTH. Second Edition, with Maps. 8vo. cloth, 10*s.* 6*d.*

MR. SHAW, M.R.C.S.

THE MEDICAL REMEMBRANCER; OR, BOOK OF EMERGENCIES: in which are concisely pointed out the Immediate Remedies to be adopted in the First Moments of Danger from Drowning, Poisoning, Apoplexy, Burns, and other Accidents; with the Tests for the Principal Poisons, and other useful Information. Fourth Edition. Edited, with Additions, by JONATHAN HUTCHINSON, F.R.C.S. 32mo. cloth, 2*s.* 6*d.*

DR. SHEA, M.D., B.A.

A MANUAL OF ANIMAL PHYSIOLOGY. With an Appendix of Questions for the B.A. London and other Examinations. With Engravings. Foolscap 8vo. cloth, 5*s.* 6*d.*

DR. SIBSON, F.R.S.

MEDICAL ANATOMY. With coloured Plates. Imperial folio. Fasciculi I. to VI. 5*s.* each.

DR. E. H. SIEVEKING.

ON EPILEPSY AND EPILEPTIFORM SEIZURES: their Causes, Pathology, and Treatment. Second Edition. Post 8vo. cloth, 10*s.* 6*d.*

MR. SINCLAIR AND DR. JOHNSTON.

PRACTICAL MIDWIFERY: Comprising an Account of 13,748 Deliveries, which occurred in the Dublin Lying-in Hospital, during a period of Seven Years. 8vo. cloth, 15*s.*

DR. SIORDET, M.B.LOND., M.R.C.P.

MENTONE IN ITS MEDICAL ASPECT. Foolscap 8vo. cloth, 2*s.* 6*d.*

MR. ALFRED SMEE, F.R.S.

GENERAL DEBILITY AND DEFECTIVE NUTRITION; their Causes, Consequences, and Treatment. Second Edition. Fcap. 8vo. cloth, 3*s.* 6*d.*

DR. SMELLIE.

OBSTETRIC PLATES: being a Selection from the more Important and Practical Illustrations contained in the Original Work. With Anatomical and Practical Directions. 8vo. cloth, 5*s.*

MR. HENRY SMITH, F.R.C.S.

I.

ON STRICTURE OF THE URETHRA. 8vo. cloth, 7*s.* 6*d.*

II.

HÆMORRHOIDS AND PROLAPSUS OF THE RECTUM: Their Pathology and Treatment, with especial reference to the use of Nitric Acid. Third Edition. Fcap. 8vo. cloth, 3*s.*

DR. J. SMITH, M.D., F.R.C.S.EDIN.

HANDBOOK OF DENTAL ANATOMY AND SURGERY, FOR THE USE OF STUDENTS AND PRACTITIONERS. Fcap. 8vo. cloth, 3*s.* 6*d.*

DR. W. TYLER SMITH.

I.

A MANUAL OF OBSTETRICS, THEORETICAL AND PRACTICAL. Illustrated with 186 Engravings. Fcap. 8vo. cloth, 12*s.* 6*d.*

II.

THE PATHOLOGY AND TREATMENT OF LEUCORRHŒA. With Engravings on Wood. 8vo. cloth, 7*s.*

DR. SNOW.

ON CHLOROFORM AND OTHER ANÆSTHETICS: THEIR ACTION AND ADMINISTRATION. Edited, with a Memoir of the Author, by Benjamin W. Richardson, M.D. 8vo. cloth, 10*s.* 6*d.*

DR. STANHOPE TEMPLEMAN SPEER.

PATHOLOGICAL CHEMISTRY, IN ITS APPLICATION TO THE PRACTICE OF MEDICINE. Translated from the French of MM. Becquerel and Rodier. 8vo. cloth, reduced to 8*s.*

MR. A. B. SQUIRE, M.B.LOND.

COLOURED PHOTOGRAPHS OF SKIN DISEASES. In Twelve Parts (one every month), with Letterpress, 3*s.* 6*d.* each.

No. I. PSORIASIS. | No. III. LICHEN. | No. V. CHLOASMA.
No. II. IMPETIGO. | No. IV. SCABIES. | No. VI. FAVUS.

MR. PETER SQUIRE.

I.

A COMPANION TO THE BRITISH PHARMACOPÆIA. Second Edition. 8vo. cloth, 8*s.* 6*d.*

II.

THE PHARMACOPÆIAS OF THIRTEEN OF THE LONDON HOSPITALS, arranged in Groups for easy Reference and Comparison. 18mo. cloth, 3*s.* 6*d.*

DR. STEGGALL.

STUDENTS' BOOKS FOR EXAMINATION.

I.

A MEDICAL MANUAL FOR APOTHECARIES' HALL AND OTHER MEDICAL BOARDS. Twelfth Edition. 12mo. cloth, 10*s.*

II.

A MANUAL FOR THE COLLEGE OF SURGEONS; intended for the Use of Candidates for Examination and Practitioners. Second Edition. 12mo. cloth, 10*s.*

III.

GREGORY'S CONSPECTUS MEDICINÆ THEORETICÆ. The First Part, containing the Original Text, with an Ordo Verborum, and Literal Translation. 12mo. cloth, 10*s.*

IV.

THE FIRST FOUR BOOKS OF CELSUS; containing the Text, Ordo Verborum, and Translation. Second Edition. 12mo. cloth, 8*s.*

V.

FIRST LINES FOR CHEMISTS AND DRUGGISTS PREPARING FOR EXAMINATION AT THE PHARMACEUTICAL SOCIETY. Second Edition. 18mo. cloth, 3*s.* 6*d.*

MR. STOWE, M.R.C.S.

A TOXICOLOGICAL CHART, exhibiting at one view the Symptoms, Treatment, and Mode of Detecting the various Poisons, Mineral, Vegetable, and Animal. To which are added, concise Directions for the Treatment of Suspended Animation. Twelfth Edition, revised. On Sheet, 2*s.*; mounted on Roller, 5*s.*

MR. FRANCIS SUTTON, F.C.S.

A SYSTEMATIC HANDBOOK OF VOLUMETRIC ANALYSIS; or, the Quantitative Estimation of Chemical Substances by Measure. With Engravings. Post 8vo. cloth, 7*s.* 6*d.*

DR. SWAYNE.

OBSTETRIC APHORISMS FOR THE USE OF STUDENTS COMMENCING MIDWIFERY PRACTICE. With Engravings on Wood. Third Edition. Fcap. 8vo. cloth, 3*s.* 6*d.*

MR. TAMPLIN, F.R.C.S.E.

LATERAL CURVATURE OF THE SPINE: its Causes, Nature, and Treatment. 8vo. cloth, 4*s.*

DR. ALEXANDER TAYLOR, F.R.S.E.

THE CLIMATE OF PAU; with a Description of the Watering Places of the Pyrenees, and of the Virtues of their respective Mineral Sources in Disease. Third Edition. Post 8vo. cloth, 7*s.*

DR. ALFRED S. TAYLOR, F.R.S.

I.

A MANUAL OF MEDICAL JURISPRUDENCE. Seventh Edition. Fcap. 8vo. cloth, 12*s.* 6*d.*

II.

ON POISONS, in relation to MEDICAL JURISPRUDENCE AND MEDICINE. Second Edition. Fcap. 8vo. cloth, 12*s.* 6*d.*

MR. TEALE.

ON AMPUTATION BY A LONG AND A SHORT RECTANGULAR FLAP. With Engravings on Wood. 8vo. cloth, 5*s.*

DR. THEOPHILUS THOMPSON, F.R.S.

CLINICAL LECTURES ON PULMONARY CONSUMPTION; with additional Chapters by E. Symes Thompson, M.D. With Plates. 8vo. cloth, 7*s.* 6*d.*

DR. THOMAS.

THE MODERN PRACTICE OF PHYSIC; exhibiting the Symptoms, Causes, Morbid Appearances, and Treatment of the Diseases of all Climates. Eleventh Edition. Revised by Algernon Frampton, M.D. 2 vols. 8vo. cloth, 28*s.*

MR. HENRY THOMPSON, F.R.C.S.

I.

STRICTURE OF THE URETHRA; its Pathology and Treatment. The Jacksonian Prize Essay for 1852. With Plates. Second Edition. 8vo. cloth, 10*s.*

II.

THE DISEASES OF THE PROSTATE; their Pathology and Treatment. Comprising a Dissertation "On the Healthy and Morbid Anatomy of the Prostate Gland;" being the Jacksonian Prize Essay for 1860. With Plates. Second Edition. 8vo. cloth, 10*s.*

III.

PRACTICAL LITHOTOMY AND LITHOTRITY; or, An Inquiry into the best Modes of removing Stone from the Bladder. With numerous Engravings, 8vo. cloth, 9*s.*

DR. THUDICHUM.

I.

A TREATISE ON THE PATHOLOGY OF THE URINE, Including a complete Guide to its Analysis. With Plates, 8vo. cloth, 14s.

II.

A TREATISE ON GALL STONES: their Chemistry, Pathology, and Treatment. With Coloured Plates. 8vo. cloth, 10s.

DR. TILT.

I.

ON UTERINE AND OVARIAN INFLAMMATION, AND ON THE PHYSIOLOGY AND DISEASES OF MENSTRUATION. Third Edition. 8vo. cloth, 12s.

II.

A HANDBOOK OF UTERINE THERAPEUTICS. Second Edition. Post 8vo. cloth, 6s.

III.

THE CHANGE OF LIFE IN HEALTH AND DISEASE: a Practical Treatise on the Nervous and other Affections incidental to Women at the Decline of Life. Second Edition. 8vo. cloth, 6s.

DR. GODWIN TIMMS.

CONSUMPTION: its True Nature and Successful Treatment. Crown 8vo. cloth, 10s.

DR. ROBERT B. TODD, F.R.S.

I.

CLINICAL LECTURES ON THE PRACTICE OF MEDICINE. *New Edition, in one Volume, Edited by* Dr. Beale, *8vo. cloth,* 18s.

II.

ON CERTAIN DISEASES OF THE URINARY ORGANS, AND ON DROPSIES. Fcap. 8vo. cloth, 6s.

MR. TOMES, F.R.S.

A MANUAL OF DENTAL SURGERY. With 208 Engravings on Wood. Fcap. 8vo. cloth, 12s. 6d.

MR. JOSEPH TOYNBEE, F.R.S., F.R.C.S.

THE DISEASES OF THE EAR: THEIR NATURE, DIAGNOSIS, AND TREATMENT. Illustrated with numerous Engravings on Wood. 8vo. cloth, 15s.

DR. TUNSTALL, M.D., M.R.C.P.

THE BATH WATERS: their Uses and Effects in the Cure and Relief of various Chronic Diseases. Third Edition. 8vo. cloth, 2s. 6d.

DR. TURNBULL.

I.

AN INQUIRY INTO THE CURABILITY OF CONSUMPTION, ITS PREVENTION, AND THE PROGRESS OF IMPROVEMENT IN THE TREATMENT. Third Edition. 8vo. cloth, 6s.

II.

A PRACTICAL TREATISE ON DISORDERS OF THE STOMACH with FERMENTATION; and on the Causes and Treatment of Indigestion, &c. 8vo. cloth, 6s.

DR. TWEEDIE, F.R.S.

CONTINUED FEVERS: THEIR DISTINCTIVE CHARACTERS, PATHOLOGY, AND TREATMENT. With Coloured Plates. 8vo. cloth, 12*s*.

VESTIGES OF THE NATURAL HISTORY OF CREATION. Eleventh Edition. Illustrated with 106 Engravings on Wood. 8vo. cloth, 7*s*. 6*d*.

DR. UNDERWOOD.

TREATISE ON THE DISEASES OF CHILDREN. Tenth Edition, with Additions and Corrections by Henry Davies, M.D. 8vo. cloth, 15*s*.

DR. UNGER.

BOTANICAL LETTERS. Translated by Dr. B. Paul. Numerous Woodcuts. Post 8vo., 2*s*. 6*d*.

MR. WADE, F.R.C.S.

STRICTURE OF THE URETHRA, ITS COMPLICATIONS AND EFFECTS; a Practical Treatise on the Nature and Treatment of those Affections. Fourth Edition. 8vo. cloth, 7*s*. 6*d*.

DR. WALKER, M.B.LOND.

ON DIPHTHERIA AND DIPHTHERITIC DISEASES. Fcap. 8vo. cloth, 3*s*.

DR. WALLER.

ELEMENTS OF PRACTICAL MIDWIFERY; or, Companion to the Lying-in Room. Fourth Edition, with Plates. Fcap. cloth, 4*s*. 6*d*.

MR. HAYNES WALTON, F.R.C.S.

SURGICAL DISEASES OF THE EYE. With Engravings on Wood. Second Edition. 8vo. cloth, 14*s*.

MR. WARING, F.R.C.S., F.L.S.

A MANUAL OF PRACTICAL THERAPEUTICS. Second Edition, Revised and Enlarged. Fcap. 8vo. cloth, 12*s*. 6*d*.

DR. WATERS, M.R.C.P.

I.

THE ANATOMY OF THE HUMAN LUNG. The Prize Essay to which the Fothergillian Gold Medal was awarded by the Medical Society of London. Post 8vo. cloth, 6*s*. 6*d*.

II.

RESEARCHES ON THE NATURE, PATHOLOGY, AND TREATMENT OF EMPHYSEMA OF THE LUNGS, AND ITS RELATIONS WITH OTHER DISEASES OF THE CHEST. With Engravings. 8vo. cloth, 5*s*.

DR. EBEN. WATSON, A.M.

ON THE TOPICAL MEDICATION OF THE LARYNX IN CERTAIN DISEASES OF THE RESPIRATORY AND VOCAL ORGANS. 8vo. cloth, 5*s*.

DR. ALLAN WEBB, F.R.C.S.L.

THE SURGEON'S READY RULES FOR OPERATIONS IN SURGERY. Royal 8vo. cloth, 10*s*. 6*d*.

DR. WEBER.

A CLINICAL HAND-BOOK OF AUSCULTATION AND PER-CUSSION. Translated by John Cockle, M.D. 5*s*.

MR. SOELBERG WELLS, M.D., M.R.C.S.

ON LONG, SHORT, AND WEAK SIGHT, and their Treatment by the Scientific Use of Spectacles. Second Edition. With Plates. 8vo. cloth, 6*s.*

MR. T. SPENCER WELLS, F.R.C.S.

I.

DISEASES OF THE OVARIES: THEIR DIAGNOSIS AND TREATMENT. Vol. I. 8vo. cloth, 9*s.*

II.

PRACTICAL OBSERVATIONS ON GOUT AND ITS COMPLICATIONS, and on the Treatment of Joints Stiffened by Gouty Deposits. Foolscap 8vo. cloth, 5*s.*

III.

SCALE OF MEDICINES WITH WHICH MERCHANT VESSELS ARE TO BE FURNISHED, by command of the Privy Council for Trade; With Observations on the Means of Preserving the Health of Seamen, &c. &c. Seventh Thousand. Fcap. 8vo. cloth, 3*s.* 6*d.*

DR. WEST.

LECTURES ON THE DISEASES OF WOMEN. Third Edition. 8vo. cloth, 16*s.*

DR. UVEDALE WEST.

ILLUSTRATIONS OF PUERPERAL DISEASES. Second Edition, enlarged. Post 8vo. cloth, 5*s.*

MR. WHEELER.

HAND-BOOK OF ANATOMY FOR STUDENTS OF THE FINE ARTS. With Engravings on Wood. Fcap. 8vo., 2*s.* 6*d.*

DR. WHITEHEAD, F.R.C.S.

ON THE TRANSMISSION FROM PARENT TO OFFSPRING OF SOME FORMS OF DISEASE, AND OF MORBID TAINTS AND TENDENCIES. Second Edition. 8vo. cloth, 10*s.* 6*d.*

DR. WILLIAMS, F.R.S.

PRINCIPLES OF MEDICINE: An Elementary View of the Causes, Nature, Treatment, Diagnosis, and Prognosis, of Disease. With brief Remarks on Hygienics, or the Preservation of Health. The Third Edition. 8vo. cloth, 15*s.*

THE WIFE'S DOMAIN: the Young Couple—the Mother—the Nurse—the Nursling. Post 8vo. cloth, 3*s.* 6*d.*

DR. J. HUME WILLIAMS.

UNSOUNDNESS OF MIND, IN ITS MEDICAL AND LEGAL CONSIDERATIONS. 8vo. cloth, 7*s.* 6*d.*

DR. WILLIAMSON, SURGEON-MAJOR, 64TH REGIMENT.

I.

MILITARY SURGERY. With Plates. 8vo. cloth, 12*s.*

II.

NOTES ON THE WOUNDED FROM THE MUTINY IN INDIA: with a Description of the Preparations of Gunshot Injuries contained in the Museum at Fort Pitt. With Lithographic Plates. 8vo. cloth, 12*s.*

MR. ERASMUS WILSON, F.R.S.

I.

THE ANATOMIST'S VADE-MECUM: A SYSTEM OF HUMAN ANATOMY. With numerous Illustrations on Wood. Eighth Edition. Foolscap 8vo. cloth, 12*s.* 6*d.*

II.

DISEASES OF THE SKIN: A Practical and Theoretical Treatise on the DIAGNOSIS, PATHOLOGY, and TREATMENT OF CUTANEOUS DISEASES. Fifth Edition. 8vo. cloth, 16*s.*

THE SAME WORK; illustrated with finely executed Engravings on Steel, accurately coloured. 8vo. cloth, 34*s.*

III.

HEALTHY SKIN: A Treatise on the Management of the Skin and Hair in relation to Health. Sixth Edition. Foolscap 8vo. 2*s.* 6*d.*

IV.

PORTRAITS OF DISEASES OF THE SKIN. Folio. Fasciculi I. to XII., completing the Work. 20*s.* each. The Entire Work, half morocco, £13.

V.

THE STUDENT'S BOOK OF CUTANEOUS MEDICINE AND DISEASES OF THE SKIN. Part I. Post 8vo. cloth, 5*s.*

VI.

ON SYPHILIS, CONSTITUTIONAL AND HEREDITARY; AND ON SYPHILITIC ERUPTIONS. With Four Coloured Plates. 8vo. cloth, 16*s.*

VII.

A THREE WEEKS' SCAMPER THROUGH THE SPAS OF GERMANY AND BELGIUM, with an Appendix on the Nature and Uses of Mineral Waters. Post 8vo. cloth, 6*s.* 6*d.*

VIII.

THE EASTERN OR TURKISH BATH: its History, Revival in Britain, and Application to the Purposes of Health. Foolscap 8vo., 2*s.*

DR. G. C. WITTSTEIN.

PRACTICAL PHARMACEUTICAL CHEMISTRY: An Explanation of Chemical and Pharmaceutical Processes, with the Methods of Testing the Purity of the Preparations, deduced from Original Experiments. Translated from the Second German Edition, by STEPHEN DARBY. 18mo. cloth, 6*s.*

DR. HENRY G. WRIGHT.

HEADACHES; their Causes and their Cure. Third Edition. Fcap. 8vo. 2*s.* 6*d.*

DR. YEARSLEY, M.D., M.R.C.S.

I.

DEAFNESS PRACTICALLY ILLUSTRATED; being an Exposition as to the Causes and Treatment of Diseases of the Ear. Sixth Edition. 8vo. cloth, 6*s.*

II.

ON THE ENLARGED TONSIL AND ELONGATED UVULA, and other Morbid Conditions of the Throat. Seventh Edition. 8vo. cloth, 5*s.*

CHURCHILL'S SERIES OF MANUALS.

Fcap. 8vo. cloth, 12*s.* 6*d.* each.

"We here give Mr. Churchill public thanks for the positive benefit conferred on the Medical Profession, by the series of beautiful and cheap Manuals which bear his imprint."—*British and Foreign Medical Review.*

AGGREGATE SALE, 137,000 COPIES.

ANATOMY. With numerous Engravings. Eighth Edition. By Erasmus Wilson, F.R.C.S., F.R.S.

BOTANY. With numerous Engravings. By Robert Bentley, F.L.S., Professor of Botany, King's College, and to the Pharmaceutical Society.

CHEMISTRY. With numerous Engravings. Ninth Edition. By George Fownes, F.R.S., H. Bence Jones, M.D., F.R.S., and A. W. Hofmann, F.R.S.

DENTAL SURGERY. With numerous Engravings. By John Tomes, F.R.S.

MATERIA MEDICA. With numerous Engravings. Fourth Edition. By J. Forbes Royle, M.D., F.R.S., and Frederick W. Headland, M.D., F.L.S.

MEDICAL JURISPRUDENCE. Seventh Edition. By Alfred Swaine Taylor, M.D., F.R.S.

PRACTICE OF MEDICINE. Second Edition. By G. Hilaro Barlow, M.D., M.A.

The MICROSCOPE and its REVELATIONS. With numerous Plates and Engravings. Third Edition. By W. B. Carpenter, M.D., F.R.S.

NATURAL PHILOSOPHY. With numerous Engravings. Fifth Edition. By Golding Bird, M.D., M.A., F.R.S., and Charles Brooke, M.B., M.A., F.R.S.

OBSTETRICS. With numerous Engravings. By W. Tyler Smith, M.D., F.R.C.P.

OPHTHALMIC MEDICINE and SURGERY. With coloured Engravings on Steel, and Illustrations on Wood. Second Edition. By T. Wharton Jones, F.R.C.S., F.R.S.

PATHOLOGICAL ANATOMY. With numerous Engravings. By C. Handfield Jones, M.B., F.R.C.P., and E. H. Sieveking, M.D., F.R.C.P.

PHYSIOLOGY. With numerous Engravings. Fourth Edition. By William B. Carpenter, M.D., F.R.S.

POISONS. Second Edition. By Alfred Swaine Taylor, M.D., F.R.S.

PRACTICAL ANATOMY. With numerous Engravings. (10*s.* 6*d.*) By Christopher Heath, F.R.C.S.

PRACTICAL SURGERY. With numerous Engravings. Fourth Edition. By William Fergusson, F.R.C.S.

THERAPEUTICS. Second Edition. By E. J. Waring, F.R.C.S., F.L.S.

Printed by W. Blanchard & Sons, 62, Millbank Street, Westminster.